Dr. V. K. CANIN

Motion Tomography

$$I_d(x_d, y_d) = t\left(\frac{x_d}{M}, \frac{y_d}{M}\right) **$$

$$= \frac{1}{4\pi d^2} \frac{1}{m^2} S\left(\frac{x_d}{m}, \frac{y_d}{m}\right) ** \frac{1}{L(k+m)^2} f\left(\frac{x_d}{k+m}, \frac{y_d}{k+m}\right)$$

$$k = -m, \text{ then}$$

$$\left(-d\frac{-z}{z}\right) \Rightarrow k = d\frac{-z}{z} \quad \text{source path}$$

$$S(x,y) \rightarrow \text{path}$$

$$k \neq -m \quad \text{blur}.$$

Linear motion tomography.

$$f(x,y) = rect\left(\frac{x}{X}\right) \delta(y)$$

Linear motion

$$I_b(x_d, y_d) = t\left(\frac{x_d}{m}, \frac{y_d}{m}\right) ** \frac{1}{4\pi d^2 m^2}\left(\varepsilon\frac{x_d}{}\right).$$

7.1

Linear tomography

$$t \Rightarrow \text{at what } z \text{ will the sinusoid disappear}$$

$$\left(\frac{1}{4\pi d^2} \frac{1}{m^2}\right)\left(a + b \cos 2\pi f_o x\right)$$

Fourier transform

Tomosynthesis $\rightsquigarrow$ source positions

Tomosynthesis

shift image by $ikx \rightarrow k = -w$
add together.

Medical Imaging Systems

PRENTICE-HALL INFORMATION AND SYSTEM SCIENCES SERIES

Thomas Kailath, *Editor*

ALBERT MACOVSKI

Professor of Electrical Engineering and Radiology
Information Systems Laboratory
Stanford University

Medical Imaging Systems

PRENTICE-HALL, INC.

Englewood Cliffs, New Jersey 07632

Library of Congress Cataloging in Publication Data

MACOVSKI, ALBERT
  Medical imaging systems.

  (Prentice-Hall information and system sciences
series)
  Bibliography: p.
  Includes index.
  1. Imaging systems in medicine.   I. Title.
II. Series. [DNLM: 1. Radiography.   2. Radionuclide
imaging.   3. Technology, Radiologic.   4. Ultrasonics—
Diagnostic use.   WN 160 M171m]
R857.06M33   1983   616.07'57   82-20465
ISBN 0-13-572685-9

Printed in the United States of America

10 9 8 7 6 5 4 3 2 1

Editorial/production supervision: Nicholas Romanelli
Cover design: Ben Santora
Manufacturing buyer: Anthony Caruso

ISBN   0-13-572685-9

Prentice-Hall International, Inc., *London*
Prentice-Hall of Australia Pty. Limited, *Sydney*
Editora Prentice-Hall do Brasil Ltda., *Rio de Janeiro*
Prentice-Hall Canada Inc., *Toronto*
Prentice-Hall of India Private Limited, *New Delhi*
Prentice-Hall of Japan, Inc., *Tokyo*
Prentice-Hall of Southeast Asia Pte. Ltd., *Singapore*
Whitehall Books Limited, *Wellington, New Zealand*

To Addie,

my source and my fulfillment

Contents

3   *Physics of Projection Radiography*   *23*

4   *Source Considerations in Radiographic Imaging*   *36*

5   *Recorder Resolution Considerations*   *63*

Preface

THIS BOOK is the outgrowth of a continuously updated set of notes for a graduate course in Electrical Engineering. The course was first offered at Stanford University in 1973 and, since then, has been offered each year.

The author has a joint appointment on the faculties of the Departments of Electrical Engineering and Diagnostic Radiology. In the Electrical Engineering Department the author is a member of the Information Systems Laboratory.

In this book we attempt to describe all the existing medical imaging systems in terms familiar to the electrical engineer. These include the impulse response, the transfer function, and the signal-to-noise ratio. In this way relatively complex systems are reduced to understandable terms, and the various design tradeoffs become clearer.

Many of the systems used have various degrees of nonlinearity or space-variant impulse responses. As such they are not capable of being structured in the elegant convolutional forms or of making use of transfer functions in the frequency domain. In each case, however, reasonable approximations are made to allow restructuring into linear invariant systems. The small loss of accuracy is more than made up by the insightful results, which are essential for the understanding and design considerations of the system.

This book is written for scientists, engineers, and graduate students interested in medical imaging. Some background in linear systems and Fourier transforms is desirable. Diffraction theory is utilized in Chapters 9 and 10 on ultrasonic imaging. However, the diffraction formulations are derived from basic principles so that no previous knowledge is assumed. The underlying physics of the imaging systems is not the main thrust of this book; however, sufficient physics is presented in each imaging modality so that the system model can be developed logically.

The mathematics used in this book is primarily the integral calculus, which dominates linear system theory. In general, impulse responses are developed for each imaging modality and then used in superposition integrals. Since the material is on imaging, multiple integrals are usually involved. Statistical considerations are utilized in evaluating signal-to-noise ratios of each modality. Here the considerations are relatively straightforward since higher-order statistics are not used. The signal-to-noise ratio is simply based on the noise variance of each picture element.

This book differs from many previous works on the physics of radiology and ultrasound in that an analytic equation is provided of the resultant image in terms of the physical parameters. Rather than verbally describing the effects of different parameters, a formal mathematical structure is provided, which should prove useful for the reader interested in further, more detailed analysis.

The author wishes to acknowledge the support and friendship of many of his colleagues and graduate students, particularly Professor William Brody and Drs. Robert Alvarez, Bruno Strul, Steve Norton, Leonard Lehmann, Dwight Nishimura and Pen-Shu Yeh.

The author also expresses his gratitude for the tireless support and encouragement of Ms. Pat Krokel, who prepared the manuscript.

A.M.

Medical Imaging Systems

1

Introduction

This text is concerned primarily with the creation of images of structures within a three-dimensional object. Although the object studied will be the human body, the knowledge developed will be applicable to a variety of nonmedical applications such as nondestructive testing.

The human body consists of tissues and organs which are primarily water, bone, and gases, with water being the dominant constituent. A wide variety of trace elements are present, such as iodine in the thyroid, tellurium in the liver, and iron in the blood. These elements play a minor role in medical imaging. This is beginning to change, especially with the advent of computerized tomography. Water, bone, and air, however, dominate the ability, or lack of it, to probe the body with various types of radiation so as to create images.

HISTORY

No effort will be made to provide a complete chronology of medical imaging. However, in an oversimplified fashion, we attempt to highlight the role of the physical scientist and engineer in some historical context.

The earliest use of techniques of this type goes back to the discovery of x-rays by William K. Røentgen in 1895. Many of the major systems' contributions to radiography, such as intensifying screens, tomography (imaging of a specific plane), and the rotating anode tube, came within the next 10 to 20 years. Thus most of the efforts in radiography, since the 1930's, have been toward improving components rather than systems.

It is interesting to note that during the latter period, profound improvements in internal visualization of disease processes were achieved by the creativity of clinicians rather than physical scientists and engineers. A variety of procedures were developed for selectively opacifying the regions of interest. These included intravenous, catheter, and orally administered dyes. Thus the radiologist, faced with the limited performance of the instrumentation, devised a variety of procedures, often invasive to the body, to facilitate visualization of otherwise invisible organs.

Beginning in the 1950's, and reaching a peak in the mid-1970's, we entered a revolutionary era in diagnostic instrumentation systems. New systems were conceived of and developed for noninvasively visualizing the anatomy and disease processes. Here the physical scientist and engineer have played the dominant roles, with clinicians being hard-pressed to keep up with the immense amount of new and exciting data.

This revolution began with nuclear medicine and ultrasound, which despite serious imaging limitations, provided noninvasive visualization of disease processes which were otherwise unavailable. The new era reached its peak with the introduction of computerized tomography in the early 1970's. Here superb cross-sectional images were obtained that rivaled the information obtained with exploratory surgery. These instruments rapidly proliferated and became the standard technique for a wide variety of procedures.

This revolutionary process is continuing, not only with profound improvements in these systems, but also with initial studies of newer speculative imaging modalities. These include the use of microwaves and nuclear magnetic resonance.

MEDICAL IMAGING MODALITIES

Present medical imaging systems in clinical use consist of three basic techniques:

1. The measurement of the transmission of x-rays through the body
2. The measurement of the reflection of ultrasonic waves transmitted through the body
3. The measurement of gamma rays emitted by radioactive pharmaceuticals which have been selectively deposited in the body

We consider each of these briefly.

TRANSMISSION OF ELECTROMAGNETIC ENERGY

It is instructive to examine the entire electromagnetic spectrum, from dc to cosmic rays, to find a region to do suitable imaging of the interior regions. The relative suitability can be evaluated based on two parameters, resolution and attenuation. To obtain a useful image, the radiation must have a wavelength under 1.0 cm in the body for resolution considerations. In addition, the radiation should be reasonably attenuated when passing through the body. If it is too highly attenuated, transmission measurements become all but impossible because of noise. If it is almost completely transmitted without attenuation, the measurement cannot be made with sufficient accuracy to be meaningful. The attenuation can be due to absorption or multiple scatter.

Figure 1.1 is an attempt to illustrate the relatively small region of the electromagnetic spectrum which is suitable for imaging of the body. In the long-wavelength region at the left, we see immediately that we have excessive attenuation at all but the very long wavelengths where the resolution would

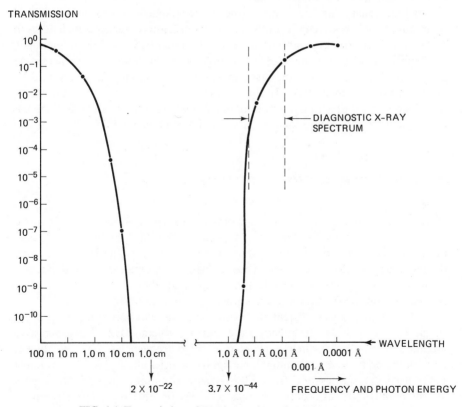

FIG. 1.1 Transmission of EM waves through 25 cm of soft tissue.

render the system unusable. Here the transmission through soft tissue can be approximated as exp $(-20l/\lambda)$, where l is the path length.

In the intermediate regions of the spectrum corresponding to the infrared, optical, and ultraviolet regions, we again have excessive attenuation due to both absorption and scatter at the myriads of tissue interfaces. This excessive attenuation continues, as shown, into the soft x-ray regions.

Between 0.5 and 10^{-2} Å, corresponding to photon energies of about 25 kev to 1.0 Mev, the attenuation is at reasonable levels with a wavelength far shorter than the resolution of interest. This is highly desirable since it ensures that diffraction will not in any way distort the imaging system and the rays will travel in straight lines. This is clearly the suitable region for imaging and represents the widely used diagnostic x-ray spectrum.

At shorter wavelengths, with the energy per photon $h\nu$ getting increasingly higher, the attenuation becomes smaller until the body becomes relatively transparent and it ceases to be a useful measurement. Also, at these shorter wavelengths, the total energy consists of relatively few quanta, resulting in poor counting statistics and a noisy image.

Before concluding our look at the electromagnetic spectrum we should point out that the wavelengths in the microwave region represent wavelengths in space. The dielectric constant of water in the microwave region is about 80, resulting in a refractive index of about 9. Thus if the body and a microwave source–detector system are immersed in water, the wavelengths are reduced by about an order of magnitude. In this case the attenuation of a 1.0-cm wave in water is no longer prohibitive and a marginal imaging system can be considered. Experiments in this regard have been made [Larsen and Jacobi, 1978].

IMAGING WITH ACOUSTIC ENERGY

Having defined the x-ray region as the only suitable part of the electromagnetic spectrum for studying the body, we now investigate the potential of acoustic radiation. The velocity of propagation of sound in water, and in most body tissues, is about 1.5×10^3 m/sec. Thus, as with electromagnetic radiation, our resolution criteria eliminate wavelengths longer than about 1.0 cm. We therefore concentrate on the frequency spectrum well above 0.15 MHz.

The attenuation coefficient in body tissues varies approximately proportional to the acoustic frequency at about 1.5 db/cm/MHz. Thus excessive attenuation rapidly becomes a problem at high frequencies. For the thicker parts of the body, as in abdominal imaging, frequencies above 5 MHz are almost never used and values of 1.0 to 3.0 MHz are common. For the imaging of shorter path lengths, as occurs in studies of the eye or other superficial structures, frequencies as high as 20 MHz can be used.

In ultrasound, unlike with x-rays, reflection images are produced, using the known velocity of propagation to calculate the depth. Unfortunately, in the frequency band where soft tissue imaging is suitable, air exhibits excessive attenuation. This is not a problem in x-rays, where air attenuation is negligible. Thus certain regions of the anatomy, primarily the lungs, cannot be studied by ultrasound. Fortunately, the entire thoracic or chest region is not covered by lungs. There is an opening in the front part of the left lung called the cardiac notch. This window allows ultrasonic studies of the heart, which are becoming increasingly important.

COMPARISON OF X-RAYS AND ULTRASOUND

It is instructive to compare the two modalities which are capable of probing the body, x-rays and ultrasound. One important distinction, which has received considerable debate, is that of toxicity. Although diagnostic x-ray levels have been considerably reduced over the years, there are considerable data indicating a small damaging effect which can increase the probability of diseases such as cancer, leukemia, and eye cataracts. The existence of a damage threshold continues to be a source of controversy. The preponderance of data at this juncture appears to indicate an absence of any toxic effects at the presently used diagnostic levels of ultrasound. These levels are well below those which produce measurable temperature changes or cavitation. This apparent toxicity advantage for ultrasound has made its use more and more popular in potentially sensitive regions such as the pregnant abdomen and the eyes.

Over and above toxicity, ultrasound and x-rays have a number of other profound differences in their imaging characteristics. Ultrasonic waves, in water, travel at about 1.5×10^3 m/sec, while x-rays have the free-space velocity of electromagnetic waves of 3×10^8 m/sec. This difference essentially means that pulse-echo or time-of-flight techniques are relatively straightforward in ultrasound and extremely difficult in x-ray systems. The propagation time in 1 cm of water is 6.7 μsec using ultrasound and 33 psec using x-rays. Thus state-of-the-art electronic techniques can quite readily resolve different depths using ultrasound but cannot begin to accomplish range gating using pulsed x-ray sources.

In addition, the velocity of propagation in the x-ray region is essentially independent of the materials used. Thus the refractive index of all structures is unity. The only mechanisms of interaction are absorption and scattering. This lack of refractive properties has both desirable and undesirable consequences. It is desirable in that the transmitted radiation travels in straight lines through the body and is undistorted by different tissue types and shapes. The undesirable aspect is that lenses cannot be constructed, so that selective imaging of specific planes is difficult. Ultrasound, on the other hand, has a wide variation in refractive indices of materials. Thus lens effects and focusing structures can be readily obtained. These same refractive effects cause some distortion within the body. Fortunately, most tissues have comparable propagation velocities, with bone and regions containing air being the primary exceptions.

Diffraction effects occur where the object of interest has structure comparable to the wavelength. In x-rays, having wavelengths less than 1 Å, these are nonexistent. However, in ultrasound the wavelength of approximately 0.5 mm can cause diffraction effects in tissue. To achieve their desired resolution properties ultrasonic imaging systems are required to operate close to the diffraction limit, whereas x-ray systems do not begin to approach this limitation.

These various properties of the radiation determine which clinical procedures they are most suitable for. For example, pulse-echo ultrasound is useful in visualizing the internal structure of the abdomen because of the distinct reflections which are received at the interfaces between organs and from various lesions within the organs. Radiological studies of the abdomen, using x-rays, show little of the internal organs because of their comparable transmission and close packing. Special procedures are often used to apply radiopaque dyes selectively to visualize specific organs. The thorax or chest, on the other hand, is essentially unavailable to ultrasound because of the air in the lungs. Radiography, however, achieves significant visualization in the chest because of the considerable differences in attenuation of air, soft tissue, and bone. Many other examples exist where the relative capabilities of the two modalities complement each other.

NUCLEAR MEDICINE

In nuclear medicine radioactive materials are administered into the body and are selectively taken up in a manner designed to indicate disease. The gamma rays emitted from these materials must be high enough in energy to escape the body without excessive attenuation. Thus higher energy gamma-ray emitters are usually desirable, the only exception being that they are more difficult to image and to detect efficiently. The energy range 25 kev to 1.0 Mev used in nuclear medicine is comparable to that of radiography, although somewhat higher on the average.

Nuclear medicine has a number of interesting features which make it very useful in diagnosis. Very small concentrations of materials are required for visualization, unlike the significant amounts of radiopaque dyes required in transmission radiography. Noninvasive intravenous administration of materials is used as compared to invasive catheterization, which is often used in radiography. In general, nuclear medicine images look poor in that they have lateral resolutions of about 1 cm and are noisy because of the limited number of photons. However, the images have the desirable characteristic of directly indicating pathology or disease processes. In many studies they are taken up only in diseased regions. Radiography, on the other hand, exhibits high resolution and relative freedom from quantum noise. However, radiographic images essentially display anatomy, so that disease processes are often distinguished by distortions of the normal anatomical features.

2

Linear Systems

DEFINITION OF LINEARITY

Many of the phenomena found in medical imaging systems exhibit *linear* behavior patterns. For example, in a nuclear medicine imaging system, where the intensity of the emitting sources double, the resultant image intensity will double. Also, if we record an image intensity due to a first source and then another due to a second source, the image intensity due to both sources acting simultaneously is the sum of the individual image intensities. These two properties, scaling and superposition, define a linear system. This can be expressed formally as

$$\mathbb{S}\{aI_1(x, y) + bI_2(x, y)\} = a\mathbb{S}\{I_1(x, y)\} + b\mathbb{S}\{I_2(x, y)\} \tag{2.1}$$

where $\mathbb{S}$ is the system operator, a and b are constants, and I_1 and I_2 are the two input functions. The functions are shown as two-dimensional since we are dealing with images.

The system operator $\mathbb{S}$ in imaging systems is typically a blurring function that smears or softens the original image. As will be discussed subsequently, $\mathbb{S}$ can be a convolution operation with the point-spread function of the system. Thus equation (2.1) is stating that the weighted sum of two blurred images is equal to the weighted sum of the two images which are then blurred by the system function. This powerful concept allows us to decompose the image, operate on the individual parts with the system function, and then sum to obtain the desired output image.

It must be emphasized that S in equation (2.1) is a linear system operator and does not apply, in general, to nonlinear systems. For example, consider a nonlinear system, such as photographic film, which exhibits a saturation value beyond which a further increase in input intensity results in no change in the recorded density. Equation (2.1) does not apply since we can have two input intensities, I_1 and I_2, each of which do not reach the saturation value, whereas their weighted sum exceeds this value. However, since the linearity property of (2.1) provides a variety of powerful techniques, we often approximate nonlinear systems into a linear form to aid in our understanding, despite the errors in the approximation. Linearization techniques of this type are often employed in electronics, where piecewise linear models are used to represent nonlinear-device characteristics.

THE SUPERPOSITION INTEGRAL, DELTA FUNCTION, AND IMPULSE RESPONSE

The linearity property expressed in equation (2.1) enables us to express the response of any linear system to an input function in an elegant and convenient manner which provides significant physical insight. We first decompose our input function into elementary functions. We then find the response to each elementary function and sum them to find the output function.

The most convenient elementary function for our decomposition is the delta function $\delta(x, y)$. The two-dimensional delta function has infinitesimal width in all dimensions and an integrated volume of unity as given by

$$\int\int_{-\infty}^{\infty} \delta(x, y)dxdy = 1. \tag{2.2}$$

The delta function can be expressed as the limit of a two-dimensional function whose volume is unity. One example is the Gaussian, as given by

$$\lim_{\alpha \to \infty} \alpha^2 \exp\left[-\pi\alpha^2(x^2 + y^2)\right] = \delta(x, y). \tag{2.3}$$

We decompose our input function $g_1(x_1, y_1)$ into an array of these two-dimensional delta functions by using the *sifting property* of the delta function, as given by

$$g_1(x_1, y_1) = \int\int_{-\infty}^{\infty} g_1(\xi, \eta)\delta(x_1 - \xi, y_1 - \eta)d\xi d\eta. \tag{2.4}$$

Here the delta function at $x_1 = \xi$, $y_1 = \eta$ has sifted out the particular value of g_1 at that point. The output function $g_2(x_2, y_2)$ is given by the system operator $\mathcal{S}$ as

$$g_2(x_2, y_2) = \mathcal{S}\{g_1(x_1, y_1)\}. \tag{2.5}$$

Using the delta function decomposition, we obtain the output function as

$$g_2(x_2, y_2) = \mathcal{S}\left\{\int\int_{-\infty}^{\infty} g_1(\xi, \eta)\delta(x_1 - \xi, y_1 - \eta)d\xi d\eta\right\}. \tag{2.6}$$

Using the linearity property of (2.1) we structure $g_1(\xi, \eta)$ as a weighting factor on each delta function so that the output becomes

$$g_2(x_2, y_2) = \int\int_{-\infty}^{\infty} g_1(\xi, \eta)\mathcal{S}\{\delta(x_1 - \xi, y_1 - \eta)\}d\xi d\eta. \tag{2.7}$$

Within the integral we have the system response to a two-dimensional delta function. This is known as the *impulse response* or *point-spread function* and is given by

$$h(x_2, y_2; \xi, \eta) = \mathcal{S}\{\delta(x_1 - \xi, y_1 - \eta)\} \tag{2.8}$$

which is the output function at x_2, y_2 due to an impulse or two-dimensional delta function at $x_1 = \xi, y_1 = \eta$. Substituting (2.8) into equation (2.7),

$$g_2(x_2, y_2) = \int\int_{-\infty}^{\infty} g_1(\xi, \eta)h(x_2, y_2; \xi, \eta)d\xi d\eta \tag{2.9}$$

we obtain the *superposition integral*, which is the fundamental concept that we derive from the linearity property (2.1). This enables us to characterize an output function completely in terms of its response to impulses. Once we know $h(x_2, y_2; \xi, \eta)$ for all input coordinates, we can find the output due to any input function g_1.

As a simple example, consider the elementary imaging system of Fig. 2.1. Here a planar array of sources $I_1(x_1, y_1)$ are separated from a surface where the output intensity $I_2(x_2, y_2)$ is recorded. As can be seen, the output is blurred due to the spreading of the radiation over the distance between planes. An impulse source at $x_1 = \xi, y_1 = \eta$, thus results in a diffuse blur having an impulse

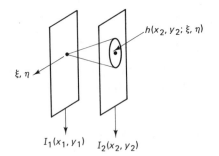

FIG. 2.1 Elementary imaging system.

response $h(x_2, y_2; \xi, \eta)$. Knowing this response for all input points enables us to calculate the output $I_2(x_2, y_2)$ for any input-source distribution $I_1(x_1, y_1)$ by using (2.9).

SPACE INVARIANCE

In many linear systems we have the added simplification that the impulse response is the same for all input points. In that case the impulse response merely shifts its position for different input points, but does not change its functional behavior. Such systems are said to be *space invariant*. The impulse response or point-spread function thus becomes dependent solely on the difference between the output coordinates and the position of the impulse as given by

$$h(x_2, y_2; \xi, \eta) = h(x_2 - \xi, y_2 - \eta). \qquad (2.10)$$

Thus, referring to Fig. 2.1, the system is space invariant if the image of the point source has the same functional form, but only translates, as the point source is moved around the x_1, y_1 plane.

The output function under these conditions becomes

$$g_2(x_2, y_2) = \int\int_{-\infty}^{\infty} g_1(\xi, \eta) h(x_2 - \xi, y_2 - \eta) d\xi d\eta. \qquad (2.11)$$

This is the two-dimensional convolution function, which is often abbreviated as

$$g_2 = g_1 ** h \qquad (2.12)$$

where the dual asterisks indicate a two-dimensional convolution. This convolutional relationship clearly indicates the "blurring" of the g_1 input function by the impulse response h. It is a very convenient form since its Fourier transform is a simple product relationship whereby the transform of the output function is the transform of the input function multiplied by the transfer function H, the transform of the impulse response. This provides an elegant relationship between the two-dimensional spatial frequencies of the input and output.

In many medical imaging systems the impulse response h varies gradually for different input coordinates. In these it becomes convenient to define regions in which this variation is negligible. These space-invariant or isoplanatic regions can be analyzed using the convenient convolutional form and thus have the benefits of having a transfer function in the Fourier transform domain.

MAGNIFICATION

A magnified image, in the general sense, is space variant since the impulse response is not dependent solely on the difference between input and output coordinates. This is seen in the pinhole imaging system shown in Fig. 2.2. With

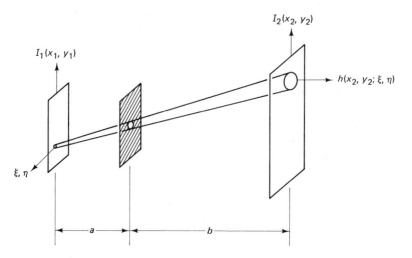

FIG. 2.2 Pinhole imaging system with magnification.

the pinhole on axis, the impulse response by geometry becomes

$$h(x_2, y_2; \xi, \eta) = h(x_2 - M\xi, y_2 - M\eta) \qquad (2.13)$$

where $M = -b/a$. Thus the output intensity is given by

$$I_2(x_2, y_2) = \iint I_1(\xi, \eta) h(x_2 - M\xi, y_2 - M\eta) d\xi d\eta. \qquad (2.14)$$

This equation can be restructured into the desirable convolutional form [Goodman, 1968] using the substitutions

$$\xi' = M\xi \quad \text{and} \quad \eta' = M\eta \qquad (2.15)$$

giving

$$I_2(x_2, y_2) = \frac{1}{M^2} \iint I_1\left(\frac{\xi'}{M}, \frac{\eta'}{M}\right) h(x_2 - \xi', y_2 - \eta') d\xi' d\eta'$$

$$= \frac{1}{M^2} I_1\left(\frac{x_2}{M}, \frac{y_2}{M}\right) ** h(x_2, y_2). \qquad (2.16)$$

Using the substitution of (2.15), we have been able to use the convolutional form. Physically, we have convolved the appropriately magnified input image with the impulse response so as to avoid the disparity between the coordinate systems. $I_1(x_2/M, y_2/M)$ is the output image we would get using an infinitesimal pinhole camera which experiences magnification but no blurring. The $1/M^2$ factor is the loss of intensity due to the magnification of the image.

The impulse response $h(x_2, y_2)$ is a magnified version of the aperture function where the magnification factor is $(a + b)/a$. In addition, the impulse response includes a collection efficiency term relating to what fraction of the intensity of each point in I_1 is collected by the aperture or pinhole. In equation (2.13) we effectively assume that the collection efficiency is a constant, independent of the x_1, y_1 coordinates, so that the aperture collects the same fraction, independent of the position of the point. If this assumption is not made, and the collection efficiency is a function of x_1 and y_1, the system becomes space variant and cannot be structured in convolutional form.

TWO-DIMENSIONAL FOURIER TRANSFORMS

Electrical and communication engineers are very familiar with Fourier analysis of various signals where a function of time is decomposed into an array of complex exponentials of the form $e^{i\omega t}$. One of the incentives for this decomposition is that the complex exponential is the eigenfunction of invariant linear systems. That is, when a complex exponential signal is applied to an invariant linear system, such as a network or filter, the same complex exponential will appear in the output with various amplitude and phase weightings. Thus, with the Fourier transform decomposition, the effect of filtering a signal is greatly facilitated by operating in the frequency domain.

In imaging systems we use the two-dimensional Fourier transform $G(u, v)$ as defined by [Bracewell, 1965]

$$G(u, v) = \mathcal{F}\{g(x, y)\} = \int\int_{-\infty}^{\infty} g(x, y) \exp\left[-i2\pi(ux + vy)\right]dxdy \qquad (2.17)$$

where $\mathcal{F}$ is the Fourier transform operator, and u and v are the spatial frequencies in the x and y dimensions. Thus the two-dimensional function is being decomposed into a continuous array of grating-like functions having different periodicities and angles. Each u, v point in the Fourier space corresponds to an elementary "plane wave" type of function in object space. This complex exponential function has lines of constant phase separated by $(u^2 + v^2)^{-1/2}$ and at an angle of $\tan^{-1}(u/v)$ with the x axis.

This decomposition into spatial frequencies u and v, having dimensions of cycles per unit distance, provides a direct measure of the spatial spectrum and bandwidth. The original function can be subject to spatial frequency filtering through various degradations in the system. We then use the inverse Fourier transform to find the resultant object function as defined by

$$g(x, y) = \mathcal{F}^{-1}\{G\} = \int\int_{-\infty}^{\infty} G(u, v) \exp\left[i2\pi(ux + vy)\right]dudv. \qquad (2.18)$$

Existence Conditions

Before using the Fourier transform we must look at the required mathematical conditions on the function $g(x, y)$ for its Fourier transform to exist. The following are the more important sufficient conditions on the function.

1. The function is absolutely integrable over the entire domain.
2. The function has only a finite number of discontinuities and a finite number of maxima and minima in any finite region.
3. The function has no infinite discontinuities.

Before becoming overly involved in these conditions, we must recall that this book relates to an applied science, so that we are interested primarily in studying the transforms of real physical phenomena. These will include a variety of two-dimensional source distributions or two-dimensional transmission functions. These physical distributions, by definition, have a transform since, as pointed out by Bracewell [1965], ". . . physical possibility is a valid sufficient condition for the existence of a transform." In our analysis, however, we will often use convenient idealized distributions to study various system properties. These include the delta function, sinusoidal distribution, dc or constant term, and so on. Each of these useful functions violates one or more of the previously listed sufficient conditions. It would be unfortunate, however, to abandon these and lose the resultant insights. Instead, we use them in a limiting process so that the conditions remain satisfied.

One example of this limiting process is the representation of a two-dimensional delta function as the limit of a two-dimensional Gaussian function as given by

$$\delta(x, y) = \lim_{\alpha \to \infty} \alpha^2 \exp(-\alpha^2 \pi r^2) \tag{2.19}$$

where $r^2 = x^2 + y^2$. The Fourier transform of this function (see Table 2.1) is given by

$$\mathcal{F}\{\alpha^2 \exp(-\alpha^2 \pi r^2)\} = \exp\left(-\frac{\pi \rho^2}{\alpha^2}\right) \tag{2.20}$$

where ρ is the radial variable in Fourier transform space and $\rho^2 = u^2 + v^2$. Applying the limiting process to equation (2.20), we obtain

$$\mathcal{F}\{\delta(x, y)\} = \lim_{\alpha \to \infty} \exp\left(-\frac{\pi \rho^2}{\alpha^2}\right) = 1. \tag{2.21}$$

Thus the Fourier transform of the delta function is uniformly distributed over the two-dimensional frequency domain.

The limiting process has allowed us to stay within the required mathematical conditions and yet evaluate the transform of a function that violates our third condition. The particular function, $\delta(x, y)$, is very useful, as is its transform. We can also use the same limiting procedure to evaluate the transform of a constant term of unity amplitude, a function that violates our first condition. In this case we define the constant term as the limit of the same Gaussian function, where α now approaches zero. Taking the limit of the transform, we evaluate the Fourier transform of unity as $\delta(u, v)$ a delta function in the frequency domain.

FOURIER TRANSFORM RELATIONSHIPS

It is important in many problems to know the relationship between manipulations in object space and those of the frequency domain. For example, what is the effect on the frequency spectrum if an object is magnified or shifted? The more important of such relationships are given below without proof. The proofs are left to the reader or by consulting references [Bracewell, 1965; Goodman, 1968].

In each of the following relationships,

$$\mathcal{F}\{g(x, y)\} = G(u, v)$$

and

$$\mathcal{F}\{h(x, y)\} = H(u, v).$$

Linearity

$$\mathcal{F}\{\alpha g + \beta h\} = \alpha \mathcal{F}\{g\} + \beta \mathcal{F}\{h\} \tag{2.22}$$

The Fourier transform operation is linear, so that the transform of the weighted sum of two functions is the weighted sum of their individual transforms.

Magnification

$$\mathcal{F}\{g(ax, by)\} = \frac{1}{|ab|} G\left(\frac{u}{a}, \frac{v}{b}\right) \tag{2.23}$$

The stretching of coordinates in one domain results in a proportional contraction in the other domain together with a constant weighting factor.

Shift

$$\mathcal{F}\{g(x - a, y - b)\} = G(u, v) \exp\left[-i2\pi(ua + vb)\right] \tag{2.24}$$

Translation of a function in object space introduces a linear phase shift in the frequency domain.

Convolution

$$\mathcal{F}\left\{\int\int_{-\infty}^{\infty} g(\xi, \eta)h(x - \xi, y - \eta)d\xi d\eta\right\} = G(u, v)H(u, v) \tag{2.25}$$

The convolution of two functions in space is represented by simply multiplying their frequency spectra. This relationship occurs very frequently in medical imaging, where a spatial distribution g is blurred by an impulse function h. The two-dimensional convolution operation is abbreviated as $g ** h$.

Cross Correlation

$$\mathcal{F}\left\{\int\int_{-\infty}^{\infty} g(\xi, \eta)h^*(x + \xi, y + \eta)d\xi d\eta\right\} = G(u, v)H^*(u, v) \qquad (2.26)$$

This relationship is abbreviated using stars as $g \star\star h^*$. It is closely related to the convolution with the equivalence expressed by

$$g(x, y) ** h(x, y) = g(x, y) \star\star h(-x, -y). \qquad (2.27)$$

If we set $g = h$, we form an *autocorrelation* where

$$\mathcal{F}\{g(x, y) \star\star g^*(x, y)\} = G(u, v)G^*(u, v) = |G(u, v)|^2. \qquad (2.28)$$

Separability

If $g(x, y)$ is separable in its rectangular coordinates as

$$g(x, y) = g_X(x)g_Y(y)$$

then

$$\mathcal{F}\{g(x, y)\} = \mathcal{F}_x\{g_X\}\mathcal{F}_Y\{g_Y\} \qquad (2.29)$$

where $\mathcal{F}_X$ and $\mathcal{F}_Y$ are one-dimensional Fourier transform operators. This relationship occurs often with simple objects and is a convenient simplification in evaluating the transforms.

A variety of interesting relationships result when the function $g(x, y)$ is separable in polar coordinates, as given by

$$g(r, \theta) = g_R(r)g_\theta(\theta). \qquad (2.30)$$

we can decompose the θ variation into its angular harmonics as

$$g_\theta(\theta) = \sum_{n=-\infty}^{\infty} a_n e^{in\theta}.$$

We then evaluate the transform of each angular harmonic by making use of the relationship

$$\mathcal{F}\{g_R(r)e^{in\theta}\} = (-i)^n e^{in\phi}\mathcal{H}_n[g_R(r)] \qquad (2.31)$$

where $\mathcal{H}_n[\cdot]$ is the *Hankel transform* of order n, which is given by

$$\mathcal{H}_n[g_R(r)] = 2\pi \int_0^\infty rg_R(r)J_n(2\pi rp)dr \qquad (2.32)$$

where p and ϕ are the Fourier transform variables in polar coordinates. Thus the complete transform is given by

$$G(p, \phi) = \mathcal{F}\{g_R(r)g_\theta(\theta)\} = \sum_{n=-\infty}^{\infty} a_n(-i)^n e^{in\phi}\mathcal{H}_n[g_R(r)] \qquad (2.33)$$

$$G(p, \phi) = 2\pi \sum_{n=-\infty}^{\infty} a_n(-i)^n e^{in\phi} \int_0^\infty rg_R(r)J_n(2\pi rp)dr. \qquad (2.34)$$

For the important case of circular symmetry, with no θ variations, we have

$$g(r, \theta) = g_R(r).$$

In this case all the a_n values are zero except for a_0, which is unity. The resultant transform becomes

$$G(\rho, \phi) = G(\rho) = 2\pi \int_0^\infty r g_R(r) J_0(2\pi r \rho) dr. \tag{2.35}$$

This Hankel transform of zero order is often referred to as the *Fourier–Bessel transform*.

FREQUENTLY OCCURRING FUNCTIONS AND THEIR TRANSFORMS

We first present the transforms of a number of well-known continuous functions (Table 2.1).

TABLE 2.1

Function	Transform
$\sin 2\pi x$	$\frac{1}{2i}[\delta(u - 1) - \delta(u + 1)]$
$\cos 2\pi y$	$\frac{1}{2}[\delta(v - 1) + \delta(v + 1)]$
$\exp[i\pi(x + y)]$	$\delta(u - \frac{1}{2}, v - \frac{1}{2})$
$\exp(-\pi r^2)$	$\exp(-\pi \rho^2)$
1	$\delta(u, v)$

We now define a number of specialized functions and present their transform without derivation (Table 2.2). Some of these are illustrated in Fig. 2.3.

TABLE 2.2

Function	Transform
$\text{rect}(x) = \begin{cases} 1 & \lvert x \rvert \leq \frac{1}{2} \\ 0 & \text{otherwise} \end{cases}$	$\text{sinc}(u) = \dfrac{\sin \pi u}{\pi u}$
$\text{rect}(x)\,\text{rect}(y)$	$\text{sinc}(u)\,\text{sinc}(v)$
$\Lambda(x) = \begin{cases} 1 - \lvert x \rvert & \lvert x \rvert \leq 1 \\ 0 & \text{otherwise} \end{cases}$	$\text{sinc}^2(u)$
$\Lambda(x)\Lambda(y)$	$\text{sinc}^2(u)\,\text{sinc}^2(v)$
$\delta(x, y) = \dfrac{\delta(r)}{\pi r}$	1
$\delta(x - x_0, y - y_0) = \dfrac{\delta(r - r_0)\delta(\theta - \theta_0)}{r}$	$e^{-i2\pi(x_0 + y_0)}$
$\text{comb}(x) = \sum\limits_{n=-\infty}^{\infty} \delta(x - n)$	$\text{comb}(u)$
$\text{comb}(x)\,\text{comb}(y)$	$\text{comb}(u)\,\text{comb}(v)$
$\text{circ}(r) = \begin{cases} 1 & r \leq 1 \\ 0 & \text{otherwise} \end{cases}$	$\dfrac{J_1(2\pi\rho)}{\rho}$
$H(x) = \begin{cases} 1 & x \geq 0 \\ 0 & \text{otherwise} \end{cases}$	$\dfrac{1}{2}\delta(u) - \dfrac{i}{2\pi u}$

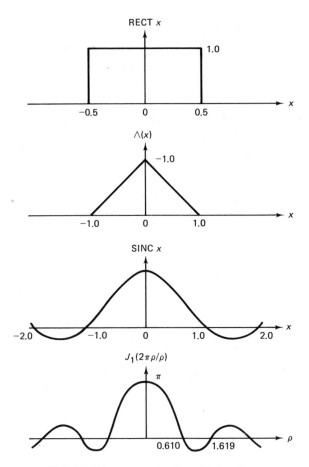

FIG. 2.3 Often encountered special functions.

SAMPLING

In many imaging operations, especially those involving computer operations, we represent a continuous two-dimensional function by a finite array of samples. We study conditions under which these samples completely define the two-dimensional function. Consider a two-dimensional image $g(x, y)$ which is sampled by a two-dimensional regular array of delta functions separated by X in the x dimension and Y in the y dimension. The resultant sampled image $g_s(x, y)$ is given by

$$g_s(x, y) = \text{comb}\left(\frac{x}{X}\right) \text{comb}\left(\frac{y}{Y}\right) g(x, y). \tag{2.36}$$

These samples will completely represent $g(x, y)$ if it is adequately bandlimited. The spectrum of the samples $g_s(x, y)$, is given by

$$G_s(u, v) = \mathcal{F}\{g_s(x, y)\} = XY \text{ comb } (uX) \text{ comb } (vY) ** G(u, v). \quad (2.37)$$

The comb functions are each infinite arrays of delta functions separated by $1/X$ and $1/Y$, respectively. Since convolution with a delta function translates a function, convolution with a comb provides an infinite array of replicated functions. Thus equation (2.37) can be rewritten as

$$G_s(u, v) = \sum_{n=-\infty}^{\infty} \sum_{m=-\infty}^{\infty} G\left(u - \frac{n}{X}, v - \frac{m}{Y}\right) \quad (2.38)$$

using the relationship

$$XY \text{ comb } (uX) \text{ comb } (uY) = \sum_{n=-\infty}^{\infty} \sum_{m=-\infty}^{\infty} \delta\left(u - \frac{n}{X}, v - \frac{m}{Y}\right). \quad (2.39)$$

$G_s(u, v)$ consists of a replicated array of the spectra $G(u, v)$. If these spectra do not overlap, the central one can be isolated through frequency-domain filtering and will thus reproduce $G(u, v)$. Since each spectra is separated by $1/X$ and $1/Y$, the filter for isolating the central replication is given by

$$H(u, v) = \text{rect } (uX) \text{ rect } (vY). \quad (2.40)$$

To avoid overlap the maximum spatial frequencies present in the image are $1/2X$ in the u axis and $1/2Y$ in the v axis. If the image is subject to this bandwidth limitation, the original spectrum can be restored using the filter of equation (2.40), where

$$G(u, v) = G_s(u, v)H(u, v). \quad (2.41)$$

Thus one approach to restoring an appropriately bandlimited image from its samples is to Fourier transform the samples, multiply the resultant spectrum by a rectangular filter, and then inverse transform the result. An equivalent method is the use of interpolation functions as derived by taking the Fourier transform of equation (2.41), where

$$g(x, y) = \left[\text{comb}\left(\frac{x}{X}\right) \text{comb}\left(\frac{y}{Y}\right)g(x, y)\right] ** h(x, y) \quad (2.42)$$

$$= XY \sum_{n=-\infty}^{\infty} \sum_{m=-\infty}^{\infty} g(nX, mY)\delta(x - nX, y - mY)$$

$$** \frac{1}{XY} \text{sinc}\left(\frac{x}{X}\right) \text{sinc}\left(\frac{y}{Y}\right) \quad (2.43)$$

$$= \sum_{n=-\infty}^{\infty} \sum_{m=-\infty}^{\infty} g(nX, mY) \text{sinc}\left[\frac{1}{X}(x - nX)\right] \text{sinc}\left[\frac{1}{Y}(x - mY)\right]. \quad (2.44)$$

The final result is quite elegant in that it is a weighted sum of two-dimensional sinc functions. Thus each sample at $x = nX$, $y = mY$ is used to weight a two-dimensional sinc function centered at that point. The sum of these functions exactly restores $g(x, y)$ if, as indicated, it is appropriately bandlimited.

The band limitation can be restated as the image having no frequency components greater than $1/2X$ and $1/2Y$, half of the sampling frequencies $1/X$

and $1/Y$. Any greater image frequency components will result in overlap of the spectral islands. This overlap cannot be removed by filtering and results in aliasing, where the higher-frequency components reappear at incorrect frequencies.

ELEMENTARY PROBABILITY

A number of considerations in medical imaging involve simple probabilities or stochastic considerations. This is especially true when studying the noise properties of the signals or images. We first define the distribution function $F(x)$ of a random variable x, which is the probability P that the outcome X of the event will be less than or equal to x as given by

$$P[X \leq x] = F(x). \tag{2.45}$$

The probability that the outcome X is between two values, x_1 and x_2, is therefore given by

$$P[x_1 \leq X \leq x_2] = F(x_2) - F(x_1) = \int_{x_1}^{x_2} p(x)dx \tag{2.46}$$

where $p(x)$ is the probability density function, which is defined as

$$p(x) = \frac{d}{dx}F(x). \tag{2.47}$$

The probability $P[\cdot]$ has the property that its value lies between zero and one. $F(x)$, the distribution function, is monotonically increasing from zero to one as x goes from $-\infty$ to ∞. The probability density function $p(x)$ is always positive and its integral over all x values is unity. The probability density function $p(x)$ takes on a variety of forms, including Gaussian, uniform, and so on. In the cases where x takes on only discrete values, such as the outcomes of dice, $p(x)$ is a sum of delta functions.

The mean of the outcomes, or expected value of X, is given by

$$\bar{X} = E(X) = \int xp(x)dx. \tag{2.48}$$

The expected value of the nth moment of X is given by substituting x^n for x in equation (2.48). An important statistical parameter is the variance σ_X^2, which represents the second central moment of X as given by

$$\sigma_X^2 = E[(X - \bar{X})^2] = E(X^2) - \bar{X}^2. \tag{2.49}$$

The standard deviation σ_X, the square root of the variance, represents the root-mean-square (rms) variation of X about the mean $\bar{X}$ as given by

$$\sigma_X = [E(X^2) - \bar{X}^2]^{1/2}. \tag{2.50}$$

If X represents the energy or power, the standard deviation is the average uncertainty in this component or the noise power. If X represents an amplitude such as voltage or current, its square represents the power and therefore its variance becomes the noise power.

In general, the signal-to-noise ratio (SNR) is defined as the ratio of signal-to-noise power. For example, in the ultrasound systems of Chapters 9 and 10, a signal $e(t)$ is received from sound reflected from a region of interest. The noise $n(t)$ from the transducer and amplifier follows a Gaussian probability density as given by

$$p(x) = \frac{1}{\sigma\sqrt{2\pi}} \exp\left[-\frac{1}{2}\left(\frac{x - \bar{x}}{\sigma}\right)^2\right]. \tag{2.51}$$

In the case of the noise signal $n(t)$, its mean is zero and its variance is σ^2, with the resultant SNR given by

$$\text{SNR} = \frac{\overline{e^2(t)}}{\sigma_n^2} \tag{2.52}$$

where σ_n^2 is the variance as shown in equation (2.51).

In the x-ray and gamma-ray portions, the noise is dominated by the discrete nature of the energy in the form of a countable number of photons recorded at each picture element. The probability density function describing the number of photons is the Poisson function given by

$$p(k) = \frac{\lambda^k e^{-\lambda}}{k!} \tag{2.53}$$

where $p(k)$ is the probability of exactly k photons where the average rate is λ photons per pixel. Applying equations (2.48) and (2.49), we find the mean $\bar{k} = \lambda$ and the variance $\sigma_k^2 = \lambda$. Since the number of photons is a unit of energy, the SNR is given by

$$\text{SNR} = \frac{\bar{k}}{\sigma_k} = \sqrt{\lambda}. \tag{2.54}$$

This is considered in detail in Chapter 6.

PROBLEMS

2.1 A pinhole imaging system of the type illustrated in Fig. 2.2 uses a circular pinhole of radius R. Using the geometry shown, and assuming a constant collection efficiency, find the output spatial frequency spectrum $I_2(u, v)$ in terms of the input spatial frequency spectrum $I_1(u, v)$.

2.2 Using the Hankel transform of zero order, often called the Fourier–Bessel transform, find the transform of the following.

(a) $\delta(r - r_0)$.

(b) $\text{rect}\left(\dfrac{r-a}{b}\right)$, where $a > b$.

(c) $g_r(ar)$, where $\mathcal{F}\{g_r(r)\} = G(\rho)$.

2.3 Prove the following properties of δ functions.

(a) $f(x, y)\, \delta(x - a, y - b) = f(a, b)\delta(x - a, y - b)$.

(b) $f(x, y) ** \delta(x - a, y - b) = f(x - a, y - b)$.

(c) $\delta(ax, by) = \dfrac{1}{|ab|}\,\delta(x, y)$.

(d) $\delta(x - x_0, y - y_0) = \dfrac{\delta(r - r_0)\delta(\theta - \theta_0)}{r}$.

(e) $\delta(x, y) = \dfrac{\delta(r)}{\pi r}$.

2.4 Prove the following Fourier transform relations.

(a) $\mathcal{F}\{\mathcal{F}\{g(x, y)\}\} = g(-x, -y)$.

(b) $\mathcal{F}\{g * h\} = GH$.

(c) $\mathcal{F}\{g(ax, by)\} = \dfrac{1}{|ab|}\,G\left(\dfrac{u}{a}, \dfrac{v}{b}\right)$.

(d) $\mathcal{F}\{g(x - a, y - b)\} = G(u, v)\exp\left[-i2\pi(ua + vb)\right]$.

2.5 A random variable x has uniform probability of being between the values a and b.

(a) Find the probability density function $p(x)$.

(b) Find the mean value of the variable $E(x)$.

(c) Find the standard deviation of the variable σ_x.

3

Physics of

Projection Radiography

The term *projection radiography* could be referred to as conventional radiography to be more familiar, although less descriptive. It refers to the bulk of x-ray studies where the transmission of x-rays through the body is recorded on film.

PARALLEL GEOMETRY

To study the various attenuation effects, we use the simplified geometry of Fig. 3.1. A parallel collimated x-ray source is assumed such as would be produced by a point source at infinity. In this assumption we avoid the distortions due to a finite source reasonably close to the object. These distortions are considered in detail in Chapter 4. This beam is partially absorbed and scattered in the object of interest with the remaining transmitted energy traveling in straight lines to the detector. For purposes of this discussion, it will be assumed that the detector plane is sufficiently far away so that all scattered radiation fails to reach the detector. The effects of scatter are considered in Chapter 6.

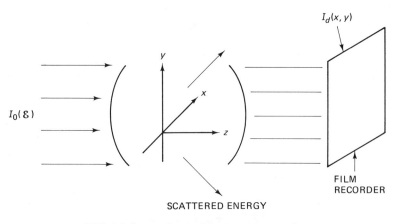

FIG. 3.1 System for studying x-ray attenuation.

ATTENUATION RELATIONSHIP

The transmitted photons either interact with a particle of matter or pass unaffected [Johns and Cunningham, 1974]. The interaction results in the removal of the photon from the beam by scattering or absorption. This interaction between an x-ray photon and a particle of matter does not affect the other photons in the beam. The number of photons interacting and removed from the beam ΔN in a region of thickness Δx is given by

$$\Delta N = -\mu N \Delta x \tag{3.1}$$

where N is the total number of impinging photons and μ is a constant of proportionality known as the *linear attenuation coefficient*. ΔN is negative since the beam loses photons. As would be expected, the number of photons interacting is proportional to the number of incident photons, the interacting distance, and the material.

If we start with N_{in} photons and, after a thickness x, have N_{out} photons, using equation (3.1) in differential form we have the integral relationship

$$\int_{N_{in}}^{N_{out}} \frac{dN}{N} = -\mu \int_{0}^{x} dx. \tag{3.2}$$

Solving equation (3.2) we have the classical attenuation relationship

$$N_{out} = N_{in} e^{-\mu x}. \tag{3.3}$$

Returning to Fig. 3.1, we use this relationship to formulate the detected intensity $I_d(x, y)$ in terms of the incident intensity I_0. The intensity attenuation follows the same relationship as that of the number of photons. The intensity is the energy per unit area, which can be expressed in terms of the number of photons per unit area weighted by the energy per photon. In the more general case the incident beam contains a spectrum of different energies, and the linear attenuation coefficient μ is a function of position and energy.

Expanding on equation (3.3), the intensity at the detector plane $I_d(x, y)$ is given by

$$I_d(x, y) = \int I_0(\mathcal{E}) \exp\left[-\int \mu(x, y, z, \mathcal{E})dz\right]d\mathcal{E} \qquad (3.4)$$

where $I_0(\mathcal{E})$ is the incident x-ray beam intensity as a function of the energy per photon $\mathcal{E}$ and $\mu(x, y, z, \mathcal{E})$ is the linear attenuation coefficient at each region of the object under study. The bracketed term represents the intensity transmission at each x, y coordinate and at each photon energy. Thus the intensity transmission t through the object at photon energy $\mathcal{E}_0$ through a thickness l at a given x, y position is given by

$$t(x, y, \mathcal{E}_0) = \exp\left[-\int_0^l \mu(x, y, z, \mathcal{E}_0)dz\right]. \qquad (3.5)$$

If μ has a uniform value of μ_0 throughout the volume, t is given by

$$t(x, y, \mathcal{E}_0) = \exp(-\mu_0 l) \qquad (3.6)$$

where μ_0 is the attenuation coefficient at $\mathcal{E}_0$.

SOURCE SPECTRUM

A number of sources of energy exist in the x-ray spectrum. However, thus far, only the x-ray tube, where an energetic electron beam strikes a metal target, has shown sufficient intensity to provide a usable image in a reasonable exposure interval. Radioactive isotopes normally have insufficient intensity, although many do provide monoenergetic radiation.

The source of x-ray energy from x-ray tubes has the Bremsstrahlung (braking) radiation spectrum [Ter-Pogossian, 1967] derived from collisional interactions between electrons and matter. This is the energy produced by the deflection and deceleration of electrons by the nucleus of the atoms in the material being bombarded. The energy is emitted in the form of x-rays or high-energy photons whose energy depends on the electron energy, the charge of the nucleus, and the distance between the electron and the nucleus. The electrons are accelerated toward the target by an anode voltage E.

In thin targets, a uniform photon energy distribution is produced whose intensity is proportional to the atomic number Z and whose maximum photon energy is the electron energy E. The spectrum is relatively uniform since different events produce different numbers of photons. For example, the photon can give up all its energy to a single emitted photon of energy E. Similarly, it can produce n photons, each of energy E/n. Thus the average intensity at each photon energy is similar. A thick target, such as is used in conventional x-ray sources, results in a triangular rather than uniform energy spectrum, as shown in Fig. 3.2. The thick target can be modeled as a sequence of thin planes each causing a successive loss of electron energy. Thus the spectrum produced by each succeeding plane is again a uniform spectrum whose maximum value becomes progressively lower. The sum of the radiation assumes a triangular form due to the reduced electron energy at increasing depths. The total emitted energy because of this triangular form is approximately proportional to ZE^2, where E is the initial electron energy.

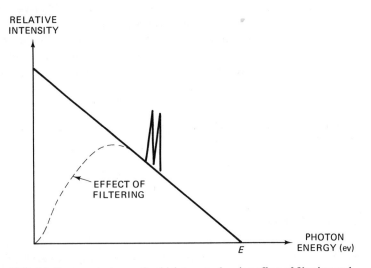

FIG. 3.2 Energy spectrum of a thick target, showing effect of filtering and characteristic radiation.

Many of the x-ray photons, especially those at lower energies, are absorbed before they leave the x-ray tube. This filtering of the beam occurs in the target itself, the glass envelope of the x-ray tube, and in other filtering structures which the x-ray beam passes through. The effect of this filtering is shown in the dashed curve in Fig. 3.2. This filtering is normally desirable since these low-energy "soft" x-rays have little penetration in the body and can cause skin burns without contributing to the transmission image. Often additional filtering material, such as aluminum, is added to remove these lower energies. In specialized studies involving relatively short path lengths and soft tissue, as in mammography, the lower-energy x-rays are used in image forming.

   Were it not for the various filtering actions within the tube, the efficiency of x-ray production would be proportional to E since the output power is proportional to E^2. Because of the filtering action, however, the output power increases at a greater rate than E^2.

   In addition to the uniformly decreasing photon energy spectrum, Fig. 3.2 shows the characteristic radiation lines. These are produced by the accelerated electrons colliding with a tightly bound electron, usually bound in the K shell of one of the atoms of the target material. The electron is ejected and the vacancy is filled by an electron from another shell. The loss of potential energy in the transition between shells is radiated as an x-ray photon having an energy equal to that of the transition. This characteristic radiation occurs at all levels but is most pronounced at the inner K shell. It represents a significant fraction of the total radiated energy from the x-ray tube.

THE ATTENUATION COEFFICIENT

The linear attenuation coefficient μ of all materials depends on the photon energy of the beam and the atomic numbers of the elements in the material [Johns and Cunningham, 1974]. Since it is the mass of the material itself that is providing the attenuation, attenuation coefficients are often characterized by μ/ρ, the mass attenuation coefficient, and are then multiplied by the density to get the linear attenuation coefficient in units of inverse distance. In the diagnostic range, below 200 kev, three mechanisms dominate the attenuation: coherent scatter, photoelectric absorption, and Compton scatter, as shown in Fig. 3.3 for water.

Coherent or Rayleigh scattering is the apparent deflection of x-ray beams caused by atoms being excited by the incident radiation and then reemitting waves at the same wavelengths. This phenomenon is useful in x-ray diffraction studies, where the x-ray energies are of the order of a few kiloelectron volts and thus the wavelengths are the same order of magnitude as atomic dimensions. It is relatively unimportant in the energies used in diagnostic radiology, as is seen by the plot of μ_r/ρ in Fig. 3.3.

At the lower energies of interest, the photoelectric effect dominates the attenuation coefficient, as seen by the plot of μ_p/ρ. The x-ray photon in this case is absorbed by interacting with a tightly bound electron. The kinetic energy of the ejected electron is dissipated in the matter. The vacancy is filled in a very short period of time by an electron falling into it, usually from the next shell. This is accompanied by the emission of characteristic x-ray photons called *fluorescent radiation*. Lower-energy excitation is absorbed in the M and L shells, while higher-energy radiation is absorbed in the inner K shell. This absorption becomes particularly important with the higher-atomic-number materials that are used as radiopaque dyes. The mass attenuation coefficient due to photoelectric absorption varies approximately as the third power of the atomic number of the material so that the linear attenuation coefficient will vary approximately as the fourth power. Thus photoelectric absorption becomes increasingly important with higher-atomic-number materials.

The attenuation coefficient undergoes a sharp increase in the energy region corresponding to the K shell. This is known as the K absorption edge. For the lower-atomic-number elements, such as are found in water and organic

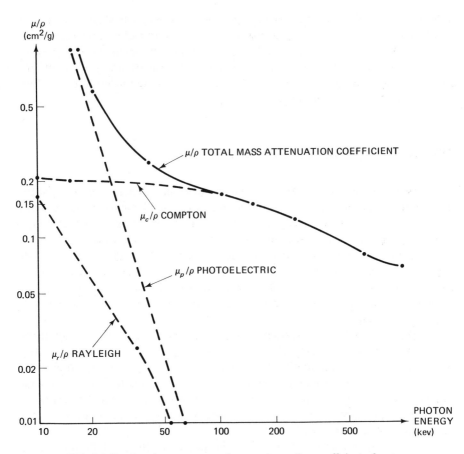

FIG. 3.3 Total and components of mass attenuation coefficient of water.

material, this K edge occurs below the diagnostic energy spectrum being used. For higher-atomic-number materials, such as lead shown in Fig. 3.4, this K edge occurs within the spectrum of interest. At energies well beyond the K absorption edge, the attenuation due to the photoelectric effect diminishes in importance. Over the range of interest, μ_p/ρ varies as $1/\mathcal{E}^3$.

The most significant and most troublesome source of tissue attenuation in the diagnostic region is that of Compton scattering. The Compton effect consists of a collision between an x-ray photon and either a free or a loosely bound electron in an outer shell, as shown in Fig. 3.5 [Weidner and Sells, 1960].

Since conservation of energy must be preserved, the relativistic energy balance is given by

$$E = E' + (m - m_0)c^2 \qquad (3.7)$$

where E' is the new photon energy and $(m - m_0)c^2$ is the increase in electron energy, where m_0 is the rest mass of the electron, c is the velocity of light, and

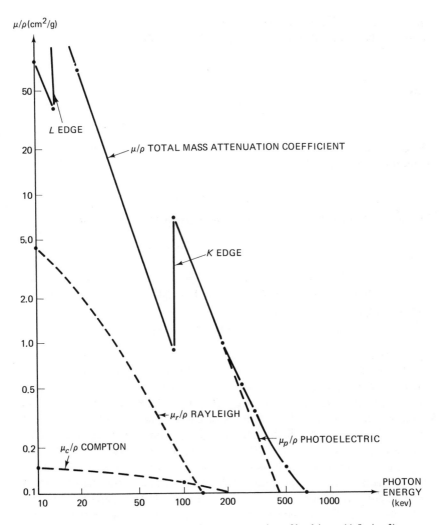

FIG. 3.4 Total and components of mass attenuation of lead ($\rho = 11.5$ g/cm²).

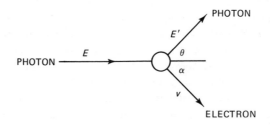

FIG. 3.5 Collision of a photon and an electron, illustrating Compton scattering.

$m = m_0/\sqrt{1 - (v/c)^2}$, the mass of the moving electron. Conservation of momentum in the x and y directions, respectively, provides

$$\left. \begin{aligned} \frac{E}{c} &= \frac{E'}{c} \cos \theta + mv \cos \alpha \\[2mm] 0 &= \frac{E'}{c} \sin \theta - mv \sin \alpha. \end{aligned} \right\} \tag{3.8}$$

and

Since the change in the energy of the photon is $E - E'$, its change in wavelength is given by

$$\Delta \lambda = \frac{hc}{E'} - \frac{hc}{E} \tag{3.9}$$

where h is Planck's constant. Using the equations above, we obtain

$$\Delta \lambda = \frac{h}{m_0 c}(1 - \cos \theta). \tag{3.10}$$

Thus the electron is scattered at an angle α and the x-ray photon is deflected by an angle θ at a lower energy (longer wavelength). The increase in wavelength $\Delta \lambda$, in angstroms, is given by

$$\Delta \lambda = 0.0241(1 - \cos \theta). \tag{3.11}$$

As can be seen, the percentage change in wavelength is significant only at relatively high energies since the wavelengths of x-rays at a photon energy of 50 kev is about 0.2 Å. Thus the scattered photons, in the diagnostic x-ray energy region, are comparable to the original energy and are a source of serious degradation to the image. The methods of eliminating the scattered photons will be discussed in Chapter 6. As seen in Fig. 3.3, Compton scattering is the largest component for water or soft tissue in most of the energy range. As would be expected, the mass attenuation coefficient due to Compton scattering varies as the electron density of the material. However, the electron densities for almost all elements are comparable at about 3×10^{23} electrons/gram. Thus the linear attenuation coefficient becomes proportional to the mass density of the material. The one exception is hydrogen, which has an electron density of about 6×10^{23} electrons/gram. Thus hydrogenous materials, such as water and soft tissue, will exhibit a proportionally higher attenuation due to Compton scattering.

As indicated by equation (3.11), the fractional change in wavelength and photon energy with angle varies significantly with the initial energy of the photon [Christensen et al., 1978]. This is illustrated in Table 3.1 for energies both within the diagnostic range and higher. As can be seen, the energy change for small angles, within the diagnostic range, is quite small.

The scattering angle distribution is approximately isotropic at the lower photon energies used in diagnostic radiology. This changes significantly at higher energies, where the scatter becomes predominantly in the forward direction. This is significant since attempts at very high energy radiography

TABLE 3.1

Incident Photon Energy (kev)	Photon Deflection Angle, θ			
	30°	60°	90°	180°
25	24.9	24.4	24	23
50	49.6	47.8	46	42
75	74.3	70	66	58
100	98.5	91	84	72
150	146	131	116	95
1000	794	508	341	205

have been limited by the relative inability to distinguish scatter from the desired transmitted radiation.

Figure 3.6 shows the various regions of the energy spectrum where the different effects dominate [Christensen et al., 1978]. As can be seen, higher-atomic-number elements will experience primarily photoelectric absorption, while those of lower atomic numbers will be dominated by Compton scattering.

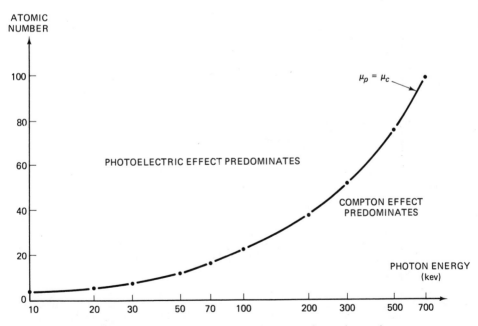

FIG. 3.6 Relative importance of the two major types of x-ray interaction. The line shows the values of Z and photon energy hv for which the photo-electric and Compton effects are equal.

Since each interaction is independent, the overall attenuation coefficient is the sum of that due to photoelectric, Rayleigh, and Compton coefficients.

In using various compounds and mixtures, a mass attenuation coefficient can be used which is given by

$$\frac{\mu}{\rho} = \sum_i w_i \frac{\mu_i}{\rho_i} \qquad (3.12)$$

where ρ is the bulk density of the material, μ_i the linear attenuation coefficient of element i, ρ_i the bulk density of element i, and w_i the fraction by weight of the element in the material.

Mass attenuation coefficients for materials normally encountered in the body are shown in Fig. 3.7. The coefficient for muscle is almost identical to that of water, while that due to fat is somewhat lower. The calcium in bone ($Z =$

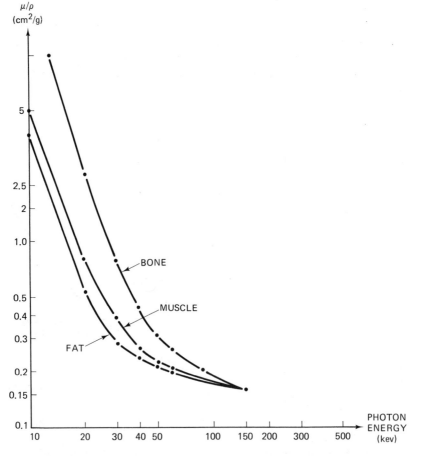

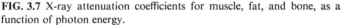

FIG. 3.7 X-ray attenuation coefficients for muscle, fat, and bone, as a function of photon energy.

20) gives it a significant photoelectric effect in the lower-energy regions, which partially accounts for its distinct visibility in radiography. At higher energies where the attenuation is primarily Compton scatter, its mass absorption coefficient becomes the same as that for other body materials.

ANALYTIC EXPRESSIONS
FOR THE ATTENUATION COEFFICIENT

The total attenuation coefficient can be decomposed into independent contributions from each mode of photon interaction as given by

$$\mu = \mu_R + \mu_P + \mu_C \tag{3.13}$$

where R, P, and C refer to Rayleigh (coherent) scattering, photoelectric effect, and Compton scattering, respectively. Rayleigh scattering is included for completeness, although, as indicated in Fig. 3.3, it plays a relatively small role in the diagnostic energy range. Thus formula (3.1), representing the total number of interacting photons, can also be used for the number of interacting photons due to each interaction mode, where μ would represent the particular μ for that mode.

A number of efforts have been made to develop reasonably accurate analytic expressions for the various components of the attenuation coefficients as a function of energy and the specific material characteristics. In general, for each element, these take the form

$$\mu = \rho N_g \left\{ f(\varepsilon) + C_R \frac{Z^k}{\varepsilon^l} + C_P \frac{Z^m}{\varepsilon^n} \right\} \tag{3.14}$$

where Z is the atomic number, ε is the photon energy in kev, C_R and C_P are constants relating the relative magnitudes of the Rayleigh and photoelectric components, $f(\varepsilon)$ is the energy-dependent Compton scattering function, and N_g is the electron mass density in electrons per gram as given by

$$N_g = N_A \frac{Z}{A} \tag{3.15}$$

where N_A is Avogadro's number and A is the atomic mass. Thus for all elements except hydrogen, N_g approximately equals $N_A/2$. The Compton scattering function $f(\varepsilon)$, which is independent of the atomic number Z, can be given to a high degree of accuracy by the Klein–Nishina function:

$$f_{KN}(\alpha) = \frac{1 + \alpha}{\alpha^2} \left[\frac{2(1 + \alpha)}{1 + 2\alpha} - \frac{1}{\alpha} \ln (1 + 2\alpha) \right] + \frac{1}{2\alpha} \ln (1 + 2\alpha) - \frac{1 + 3\alpha}{(1 + 2\alpha)^2} \tag{3.16}$$

where $\alpha = \varepsilon/510.975$ kev. In the energy range of interest, a simpler function having reasonable accuracy is $f(\varepsilon) = 0.597 \times 10^{-24} \exp [-0.0028(\varepsilon - 30)]$.

The exponents in the Rayleigh and photoelectric components have been experimentally determined as $k = 2.0$, $l = 1.9$, $m = 3.8$, and $n = 3.2$, and the constants as $C_R = 1.25 \times 10^{-24}$ and $C_P = 9.8 \times 10^{-24}$.

For composite materials the attenuation coefficient becomes

$$\mu = \rho N_g \left\{ f(\mathcal{E}) + C_R \frac{\bar{Z}_r^k}{\mathcal{E}^l} + C_P \frac{\bar{Z}_p^m}{\mathcal{E}^n} \right\} \tag{3.17}$$

where $\bar{Z}_r$ and $\bar{Z}_p$ are the effective atomic numbers as given by

$$\bar{Z}_r = \left(\sum_i \alpha_i Z_i^k \right)^{1/k} \tag{3.18}$$

and

$$\bar{Z}_p = \left(\sum_i \alpha_i Z_i^m \right)^{1/m} \tag{3.19}$$

where α_i is the electron fraction of the ith element given by

$$\alpha_i = \frac{N_{gi}}{\sum_j N_{gj}} \tag{3.20}$$

where $N_{gi} = N_A w_i (Z_i/A_i)$. In this composite material N_g becomes

$$N_g = \sum N_{gi} = N_A \sum_i w_i \frac{Z_i}{A_i}. \tag{3.21}$$

Thus, as previously discussed, the important attenuation mechanisms in the diagnostic energy range are a photoelectric component having a very strong Z dependence which dominates the lower energies, and a Z independent Compton scattering component which dominates the higher energies.

PROBLEMS

3.1 A region of the body has a 10-cm thickness of muscle tissue. Part of this region has a superimposed 2-cm thickness of bone. The densities of the muscle and bone are 1.0 and 1.75, respectively. Calculate the x-ray transmission in the muscle tissue alone and in the combined muscle and bone region at energies of 30 and 100 kev. Which energy is preferable for bone–muscle contrast? Which is preferable for visualizing variations in muscle thickness in the presence of bone?

3.2 Using the simpler analytic expression for the attenuation coefficient, in the region where the photon energy is under 100 kev, find an approximate expression for the energy at which the photoelectric effect and Compton effect are equal. Check your expression with that of Fig. 3.6 at $Z = 10$ and 20.

3.3 A source emits photons of energy E to a thin object as shown in Fig. P3.3.

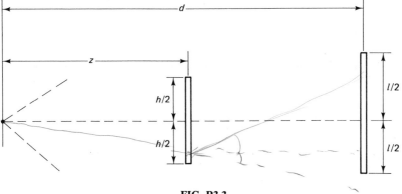

FIG. P3.3

(a) Find the minimum energy of a Compton-scattered photon reaching the x-ray detector assuming single scattering events.

(b) Repeat part (a) where $E = 100$ kev, $z = d/2$, $h = d/4$, and $l = d/4$.

3.4 Consider a cube of depth l having a total of N incident photons on one face. Calculate the total number of scattered photons if the total linear attenuation coefficient is μ and the Comptom coefficient is μ_c. [*Hint*: Use the general relationship for the number of interactions in a thin section, $\Delta n = n_{\text{in}} \mu \Delta x$.] Neglect multiple scattering.

3.5 X-ray transmission measurements are made of a single material of known length at two photon energies in the diagnostic energy range where only the Compton and photoelectric components are significant. What information can be derived about the material if it is known that it is

(a) an element?

(b) a compound or mixture?

4

Source Considerations
in Radiographic Imaging

In Chapter 3, which emphasized attenuation mechanisms, a parallel x-ray source was assumed. In this chapter we study the limits imposed by an x-ray source of finite size.

POINT-SOURCE GEOMETRY

In radiography the sources approach point sources resulting from an electron beam striking a metal target. The use of a point source, with its associated diverging beam, results in "distorted" projection images compared to those of the parallel beam studied in Chapter 3 [Christensen et al., 1978]. A typical x-ray tube is shown in Fig. 4.1 [Ter-Pogossian, 1967]. The electron beam, accelerated to about 100 kv, is used to bombard a tungsten anode. Since the exposure times are a small fraction of a second, the anode heating is minimized by using a rotating anode and thus providing a larger dissipation surface. The electron

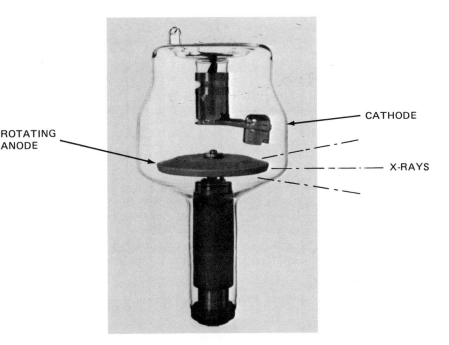

FIG. 4.1 Rotating anode x-ray tube. (Courtesy of the Machlett Laboratories, Inc.)

beam strikes a tilted surface so that the projected focal spot, in the direction of the beam, is smaller than the bombarded area.

We first consider the geometry formed by an ideal point source as shown in Fig. 4.2. The output is formed by the line integral of the attenuation coefficient $\mu(x, y, z)$ of the various rays. In studying the geometric considerations relating to image distortion and resolution, it is convenient to assume a monoenergetic source. This represents no loss of generality since we can always return to the general relationship as expressed in equation (3.4). Thus the detector output due to a monoenergetic source is given by

$$I_d(x_d, y_d) = I_i(x_d, y_d) \exp\left[- \int \mu_0(x, y, z)dr \right] \tag{4.1}$$

where $I_i(x_d, y_d)$ is the intensity incident on the detector plane in the absence of any attenuation and $\mu_0(x, y, z)$ is the linear attenuation coefficient at the monochromatic energy $\mathcal{E}_0$.

This intensity $I_i(x_d, y_d)$ in the absence of any attenuating object can be evaluated with the aid of Fig. 4.3. A point radiator emits N photons isotropically during the exposure interval. The intensity at a point x_d, y_d in the detector plane is proportional to the number of photons per unit area at that point as

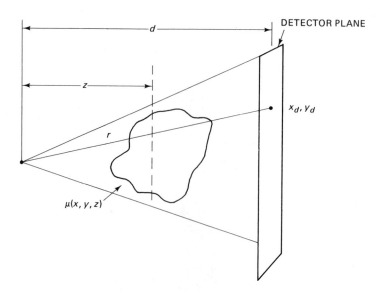

FIG. 4.2 Point-source x-ray system.

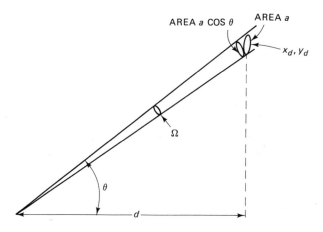

FIG. 4.3 Intensity falloff of an incident beam.

given by

$$I_i(x_d, y_d) = K\frac{N\Omega}{4\pi a} \tag{4.2}$$

where $N\Omega/4\pi$ is the number of photons in Ω, a is the incremental area, K is a constant representing the energy per photon, and Ω is the solid angle intercepted by the area a given by

$$\Omega = \frac{a\cos\theta}{r^2}. \tag{4.3}$$

It is often convenient to specify $I_i(x_d, y_d)$ in terms of I_0, its value at the origin, where $\theta = 0$, as given by

$$I_0 = \frac{KN}{4\pi d^2}.\tag{4.4}$$

This provides a representation that shows the variations in incident intensity with detector coordinates. Thus I_i can be expressed as

$$I_i = I_0 \cos^3 \theta = I_0 \frac{1}{(1 + r_d^2/d^2)^{3/2}}\tag{4.5}$$

where $r_d^2 = x_d^2 + y_d^2$.

This $\cos^3 \theta$ dependence can be interpreted as the product of an inverse square falloff with distance, providing a $\cos^2 \theta$ dependence, multiplied by a $\cos \theta$ dependence due to the obliquity between the rays and the detector plane. Thus far the source has been assumed to be monoenergetic. For a polychromatic source the detector output becomes

$$I_d(x_d, y_d) = \int I_i(\mathcal{E}) \exp\left[- \int \mu(x, y, z, \mathcal{E})dr \right] d\mathcal{E}.\tag{4.6}$$

DEPTH-DEPENDENT MAGNIFICATION

Using equation (4.1), the simplified monoenergetic case, we can develop a more useful formulation which directly illustrates the "distortion" due to point-source geometry. The line integral element, dr, is decomposed as

$$dr = \sqrt{dx^2 + dy^2 + dz^2}.\tag{4.7}$$

The line integration, as seen in Fig. 4.2, takes place along a line defined as

$$x = \frac{x_d}{d}z \quad \text{and} \quad y = \frac{y_d}{d}z.\tag{4.8}$$

These equations allow us to rewrite the line integration of equation (4.1) in terms of the depth z using

$$dr = dz\sqrt{1 + \left(\frac{dx}{dz}\right)^2 + \left(\frac{dy}{dz}\right)^2}$$

$$\frac{dx}{dz} = \frac{x_d}{d} \quad \text{and} \quad \frac{dy}{dz} = \frac{y_d}{d}.$$

Substituting, we obtain

$$I_d(x_d, y_d) = I_i \exp\left[-\sqrt{1 + \frac{r_d^2}{d^2}} \int \mu_0\left(\frac{x_d}{d}z, \frac{y_d}{d}z, z\right)dz \right].\tag{4.9}$$

The two-dimensional transmission function at any plane z is magnified by d/z in the detector plane, as can be seen in Fig. 4.2. We can therefore rewrite the preceding equation as

$$I_d(x_d, y_d) = I_i \exp\left[-\sqrt{1 + \frac{r_d^2}{d^2}} \int_0^d \mu_0\left(\frac{x_d}{M(z)}, \frac{y_d}{M(z)}, z\right)dz\right] \quad (4.10)$$

where $M(z) = d/z$. This formulation can be arrived at through physical and geometric reasoning. The radical outside the integral is the obliquity factor due to the longer path lengths of rays through the object at greater angles to the normal. In certain geometries, as with relatively thin objects, it can be ignored.

EXAMPLES OF POINT-SOURCE GEOMETRY

As a first example, in Fig. 4.4 we study an infinite slab of thickness L which is centered at a depth of z_0 and has a uniform attenuation of μ_a. The three-dimensional attenuation coefficient can be expressed as

$$\mu_0(x, y, z) = \mu_a \operatorname{rect}\left(\frac{z - z_0}{L}\right). \quad (4.11)$$

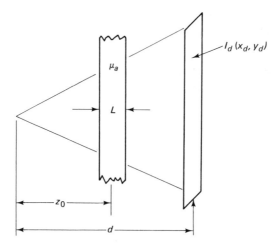

FIG. 4.4 Imaging of an infinite slab.

Since μ_0 is a function of z only, equation (4.10) simply involves the integral of a rect function, giving

$$I_d(x_d, y_d) = I_i \exp\left(-\sqrt{1 + \frac{r_d^2}{d^2}}\mu_a L\right). \quad (4.12)$$

For the case where $(r_d^2/d^2)\mu_a L \ll 1$, corresponding to a combination of a small attenuation coefficient, thin section, or regions close to the axis, the detected output can be approximated by

$$I_d(x_d, y_d) \simeq I_i e^{-\mu_a L}.$$

Another interesting example of the effects of point-source geometry is the imaging of a rectangular object of unlimited extent in the x direction as shown in Fig. 4.5. The attenuation coefficient in space is defined by

$$\mu_0(x, y, z) = \mu_a \, \text{rect}\left(\frac{y}{L}\right) \text{rect}\left(\frac{z - z_0}{w}\right) \qquad (4.13)$$

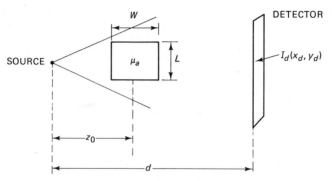

FIG. 4.5 Imaging of a rectangular object.

where μ_a is the uniform value of μ_0 throughout the object. The resultant intensity pattern at the detector becomes

$$I_d(x_d, y_d) = I_i \exp\left[-\sqrt{1 + \frac{r_d^2}{d^2}} \int \mu_a \, \text{rect}\left(\frac{y_d z}{dL}\right) \text{rect}\left(\frac{z - z_0}{w}\right) dz \right]. \qquad (4.14)$$

The product of the two rect functions is used to define the upper and lower limits of integration corresponding to the overlap region of the two functions, as shown in Fig. 4.6.

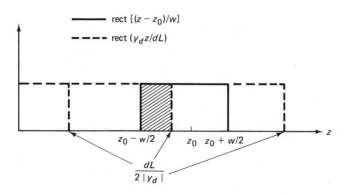

FIG. 4.6 Product of rect functions.

Rect $[(z - z_0)/w]$ is shown centered at z_0 with a width w. Three cases are shown for rect $y_d z/dL$, corresponding to three ranges of y_d. Since the function is symmetrical in y_d we can evaluate it for $|y_d|$, with the same image appearing at each side of the y_d axis. For values of $|y_d| > dL/(2z_0 - w)$, the rect functions do not overlap, providing an upper and lower limit of integration of $dL/2|y_d|$. Thus the integrated value is zero, corresponding to the lack of attenuation in the region where the rays miss the object. In the next region, for values of $|y_d|$ below $dL/(2z_0 - w)$ but above $dL/(2z_0 + w)$, the integration takes place in the shaded region from $z_0 - w/2$ to $dL/2|y_d|$, corresponding to rays cutting through the corners of the object. In the third region, where $|y_d| < dL/(2z_0 + w)$, the rays always go through the entire object. This corresponds to an integrated value of w since rect $(z - z_0)/w$ determines the limits of integration. The resultant equation is given by

$$I_d(x_d, y_d) = I_i \exp\left(-\mu_a\sqrt{1 + \frac{r_d^2}{d^2}} \int_{\min\left\{\frac{z_0 - w/2}{\frac{dL}{2|y_d|}}\right\}}^{\min\left\{\frac{z_0 + w/2}{\frac{dL}{2|y_d|}}\right\}} dz\right). \qquad (4.15)$$

Figure 4.7 illustrates the transmission I_d/I_i versus $|y_d|$, ignoring the obliquity factor $\sqrt{1 + r_d^2/d^2}$.

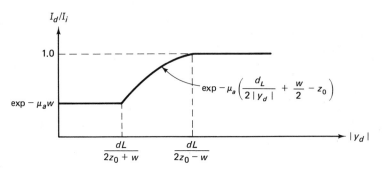

FIG. 4.7 Transmission of a rectangular object.

For very thin sections the attenuation coefficient can be characterized as

$$\mu_0(x, y, z) = \tau(x, y)\delta(z - z_0) \qquad (4.16)$$

where $\tau(x, y)$ is the line integral of the attenuation coefficient at each point x, y. The resulting detected output is given by

$$I_d(x_d, y_d) = I_i \exp\left[-\sqrt{1 + \frac{r_d^2}{d^2}}\tau\left(\frac{x_d}{M}, \frac{y_d}{M}\right)\right] \qquad (4.17)$$

where $M = d/z_0$. If we ignore the obliquity factor $\sqrt{1 + r_d^2/d^2}$, we can rewrite the detected output as

$$I_d(x_d, y_d) = I_i t\left(\frac{x_d}{M}, \frac{y_d}{M}\right) \tag{4.18}$$

where the transmission function $t(x, y) = \exp\left[-\tau(x, y)\right]$.

In this form the geometric magnification factor of the diverging beam is evident where a plane at z_0 is magnified an amount d/z_0 in the detector plane. Planes close to the source receive a large magnification, while those close to the detector plane receive a magnification approaching unity.

The distortion of images due to point-source geometry can cause significant problems in clinical interpretation if the diagnostician does not take it into account [Christensen et al., 1978]. For example, in Fig. 4.8 the apparent relative

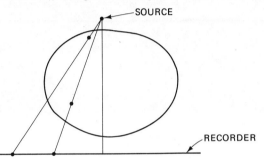

FIG. 4.8 Distortion of the relative position of two images with respect to the center lines.

radial position of two objects is distorted. Similarly, in Fig. 4.9, the apparent size of a tilted object depends on its position within the diverging beam. These examples are exaggerated compared to the usual clinical situation where the angular divergence of the beam is relatively small.

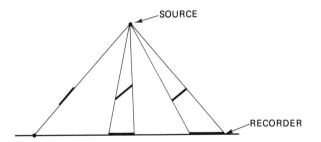

FIG. 4.9 Apparent size of a tilted object varies with its lateral position.

Figure 4.10 illustrates an x-ray photograph of an off-axis elongated plastic hollow cylinder. Note how the variation in magnification with depth gives the appearance of a truncated conical section, with one end experiencing greater magnification than the other.

FIG. 4.10 X-ray photograph of a hollow plastic cylinder.

EXTENDED SOURCES

We have shown how the use of a point x-ray source produces images having a depth-dependent magnification which is a distortion when compared to the parallel geometry of Fig. 3.1. We now consider the effects of a finite source.

The x-ray source using a bombarded target, as in Fig. 4.1, has finite dimensions, which significantly affects the resolution of the detected image [Sprawls, 1977]. We first assume that the source is planar and parallel to the detector plane as shown in Fig. 4.11. If the object is an opaque plane at z having an array of pinholes, each pinhole will reproduce an inverted image of the source magnified by $(d - z)/z$, as indicated by the geometry. The point response $h(x_d, y_d)$ for a pinhole at the origin in plane z, for a source distribution $s(x_s, y_s)$, is of the form

$$h(x_d, y_d) = Ks\left(-x_d\frac{z}{d - z}, -y_d\frac{z}{d - z}\right) \tag{4.19}$$

$$= Ks\left(\frac{x_d}{m}, \frac{y_d}{m}\right) \tag{4.20}$$

where K is a proportionality constant and m, the magnification of the source due to a hole in plane z, is given by

$$m(z) = -\frac{d - z}{z} = 1 - M(z). \tag{4.21}$$

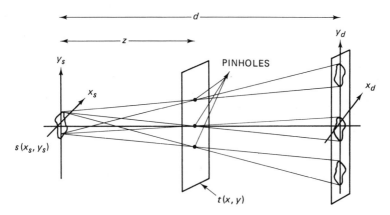

FIG. 4.11 Planar extended source.

Thus the magnification of the source image is 1 minus the magnification of the object. Since the response due to each pinhole is independent of its lateral position, the system is space invariant, as discussed in Chapter 2. The response to each isolated plane can be structured in convolution form and the spatial frequency domain can conveniently be used. The response to a transparency having transmission $t(x, y)$ in plane z is given by

$$I_d(x_d, y_d) = Kt\left(\frac{x_d}{M}, \frac{y_d}{M}\right) ** s\left(\frac{x_d}{m}, \frac{y_d}{m}\right). \qquad (4.22)$$

This can be expressed in the Fourier domain as the product of the individual transforms, where

$$I_d(u, v) = KM^2m^2T(Mu, Mv)S(mu, mv) \qquad (4.23)$$

where T and S are the Fourier transforms of t and s, and u and v are the spatial frequency coordinates.

ANALYSIS OF IMAGING USING PLANAR SOURCES

Having explored a simplified view of the effects of extended sources, we now formulate a more general analysis. Figure 4.12 illustrates an imaging system using a planar source $s(x_s, y_s)$. In our analysis, we first find the detected image due to a differential point at x_s, y_s on the source distribution. We then find the total detected intensity $I_d(x_d, y_d)$ by integrating over the entire source. The differential intensity at the detector plane in the absence of the object, $dI_i(x_d, y_d)$, due to a point at x_s, y_s is again given by

$$dI_i(x_d, y_d) = dI_0 \cos^3 \theta$$

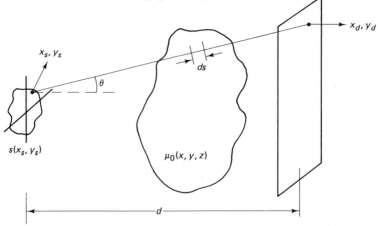

FIG. 4.12 Imaging, using a planar source.

as in (4.5). dI_0 is now defined as the differential detected intensity on the axis of the particular infinitesimal source point where $x_d = x_s$ and $y_d = y_s$, as given by

$$dI_0 = \frac{s(x_s, y_s)dx_sdy_s}{4\pi d^2}. \tag{4.24}$$

The angular distribution $\cos^3 \theta$ is also measured from each infinitesimal source point as given by

$$\cos^3 \theta = \left[1 + \left(\frac{x_d - x_s}{d}\right)^2 + \left(\frac{y_d - y_s}{d}\right)^2\right]^{-3/2}$$

$$= \frac{1}{(1 + r_{ds}^2/d^2)^{3/2}} \tag{4.25}$$

where $r_{ds} = [(x_d - x_s)^2 + (y_d - y_s)^2]^{1/2}$, the lateral distance between source and detector points.

Inserting the object with attenuation $\mu_0(x, y, z)$, the differential detected intensity due to each infinitesimal source point is given by

$$dI_d(x_d, y_d, x_s, y_s) = \frac{s(x_s, y_s)dx_sdy_s}{4\pi d^2(1 + r_{ds}^2/d^2)^{3/2}} \exp\left[-\int \mu_0(x, y, z)ds\right]$$

$$= dI_i \exp\left[-\int \mu_0(x, y, z)ds\right] \tag{4.26}$$

where ds is the element of line integration. Expanding ds, we obtain

$$ds = \sqrt{dx^2 + dy^2 + dz^2}$$

$$= dz\sqrt{1 + \left(\frac{dx}{dz}\right)^2 + \left(\frac{dy}{dz}\right)^2}. \tag{4.27}$$

Again parameterizing x and y coordinates in terms of z, the line integration takes place along

$$x = \frac{x_d - x_s}{d}z + x_s \quad \text{and} \quad y = \frac{y_d - y_s}{d}z + y_s. \tag{4.28}$$

Substituting equations (4.27) and (4.28) into (4.26), we obtain

$$dI_d(x_d, y_d, x_s, y_s) = dI_i \exp\left[-\sqrt{1 + \frac{r_{ds}^2}{d^2}} \int \mu_0\left(\frac{x_d - x_s}{d}z + x_s,\right.\right.$$

$$\left.\left.\frac{y_d - y_s}{d}z + y_s, z\right)dz\right]. \tag{4.29}$$

Using the previously defined object magnification $M = d/z$ and source magnification $m = -(d - z)/z$, (4.29) becomes

$$dI_d(x_d, y_d, x_s, y_s) = dI_i \exp\left[-\sqrt{1 + \frac{r_{ds}^2}{d^2}} \int \mu_0\left(\frac{x_d - mx_s}{M}, \frac{y_d - my_s}{M}, z\right)dz\right]. \tag{4.30}$$

The detected image due to the entire source, $I_d(x_d, y_d)$, is obtained by integrating the image due to a source point $dI_d(x_d, y_d, x_s, y_s)$ over the entire source as given by

$$I_d(x_d, y_d) = \int\int dI_d(x_d, y_d, x_s, , y_s)$$

$$= \frac{1}{4\pi d^2}\int\int \frac{s(x_s, y_s)}{(1 + r_{ds}^2/d^2)^{3/2}} \exp\left[-\sqrt{1 + \frac{r_{ds}^2}{d^2}} \int \mu_0\left(\frac{x_d - mx_s}{M},\right.\right.$$

$$\left.\left.\frac{y_d - my_s}{M}, z\right)dz\right] dx_s dy_s. \tag{4.31}$$

For the more complete polyenergetic case, both s and μ are functions of energy and the entire expression is integrated over the energy spectrum.

We can simplify equation (4.31) to provide more insight into the imaging process. We first assume that r_{ds} is sufficiently smaller than d so that we can ignore the two obliquity factors relating to the falloff in source intensity and the increased path through the object. We study a thin object at $z = z_0$ again characterized by

$$\mu_0(x, y, z) = \tau(x, y)\delta(z - z_0). \tag{4.32}$$

The resultant detected image intensity becomes

$$I_d(x_d, y_d) = \frac{1}{4\pi d^2}\int\int s(x_s, y_s) \exp\left[-\tau\left(\frac{x_d - mx_s}{M}, \frac{y_d - my_s}{M}\right)\right]dx_s dy_s. \tag{4.33}$$

To place this expression in the desired space-invariant convolutional form, we use the substitution

$$x_s' = mx_s, \quad y_s' = my_s \tag{4.34}$$

to provide

$$I_d(x_d, y_d) = \frac{1}{4\pi d^2 m^2} s\left(\frac{x_d}{m}, \frac{y_d}{m}\right) ** \exp\left[-\tau\left(\frac{x_d}{M}, \frac{y_d}{M}\right)\right]. \tag{4.35}$$

Using the previously defined $t(x, y) = \exp[-\tau(x, y)]$, we obtain the simplified convolution expression

$$I_d(x_d, y_d) = \frac{1}{4\pi d^2 m^2} s\left(\frac{x_d}{m}, \frac{y_d}{m}\right) ** t\left(\frac{x_d}{M}, \frac{y_d}{M}\right) \tag{4.36}$$

which is identical to (4.22). This result was derived using a superposition of source points each separately imaging the object. This provided a general result, equation (4.31), for imaging any object. An alternative, simpler approach, however, can be used for the case of a planar object with the obliquity factors ignored, giving the same result as in equation (4.36).

ALTERNATIVE ANALYSIS USING PLANAR OBJECTS

In this approach we find the detected intensity from the entire source due to a transparency consisting of an impulse where $t(x, y) = \delta(x - x', y - y')$, as shown in Fig. 4.13. The resultant intensity at the detector plane or impulse response $h(x_d, y_d, x', y')$ is given by

$$h(x_d, y_d, x', y') = \frac{\eta}{m^2} s\left(\frac{x_d - Mx'}{m}, \frac{y_d - My'}{m}\right) \tag{4.37}$$

where η is the collection efficiency of the pinhole as given by

$$\eta = \frac{\Omega}{4\pi} \tag{4.38}$$

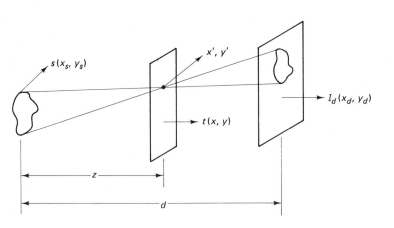

FIG. 4.13 Impulse response with an extended source.

where Ω is the solid collection angle of the pinhole. Equation (4.37) is derived by direct geometric projection with a magnification m and a translation weighted by M. The term η/m^2 is the collection efficiency divided by the ratio of image and source areas. Ignoring obliquity is equivalent to assuming that the solid angle of the unity area pinhole is $1/z^2$ over the entire transparency. With this approximation, the detected intensity is given by

$$I(x_d, y_d) = \int\int h(x_d, y_d, x', y')t(x', y')dx'dy'$$

$$= \frac{1}{4\pi z^2 m^2} \int\int t(x', y')s\left(\frac{x_d - Mx'}{m}, \frac{y_d - My'}{m}\right)dx'dy'. \quad (4.39)$$

Substituting $x'' = Mx'$ and $y'' = My'$ provides a convolution relationship given by

$$I_d(x_d, y_d) = \frac{1}{4\pi z^2 m^2 M^2}t\left(\frac{x_d}{M}, \frac{y_d}{M}\right) ** s\left(\frac{x_d}{m}, \frac{y_d}{m}\right) \quad (4.40)$$

which is identical to the previously derived equation (4.36).

Effects of Source Size

Equation (4.40) illustrates the basic problem of the loss of resolution due to source size. For object planes close to the detector where $M \simeq 1$ and $m \simeq 0$, the image has unity magnification and is not blurred by the source, no matter what its size, since $(1/m^2)s(x/m, y/m)$ approaches a delta function. For object planes closer to the source, for example at $z = d/2$ where $M = 2$ and $|m| = 1$, the object plane will be blurred by the source size itself. Attempts at greater magnifications will have greater blurring since $|m| = M - 1$. Figure 4.14 illustrates two x-ray photographs of a high-resolution test object taken with different magnifications. In the case of unity magnification the array of holes are well resolved due to the lack of blur from the source function. With a magnification of 2, however, the smaller holes are clearly blurred by the source function.

SIMPLIFYING RELATIONSHIPS USING SOLID OBJECTS

The simplified convolution relationships (4.36) and (4.40) were derived for a planar object with the only approximations being the neglecting of obliquity factors. However, for the solid object, even with the neglecting of obliquity, the nonlinear relationship prevents us from forming a convolution relationship.

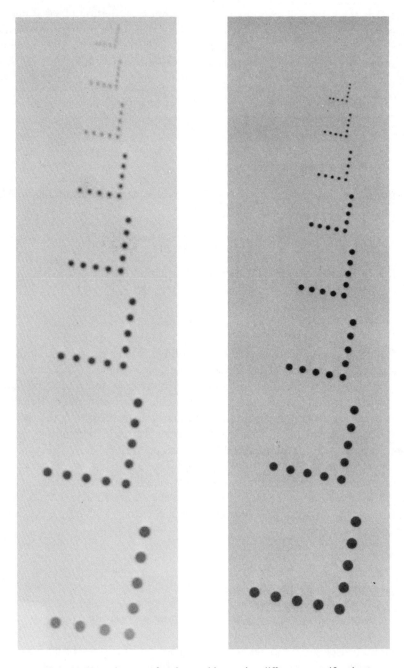

FIG. 4.14 X-ray images of a planar object, using different magnifications.

Repeating (4.31) without the obliquity factors, we have

$$I_d(x_d, y_d) = \frac{1}{4\pi d^2} \int\int s(x_s, y_s) \exp\left[-\int \mu_0\left(\frac{x_d - mx_s}{M}, \frac{y_d - my_s}{M}, z\right)dz \right]dx_s dy_s.$$

$$(4.41)$$

In general the three-dimensional attenuation coefficient of the object $\mu(x, y, z)$ must be used to solve for the intensity. Unfortunately, this relationship does not provide the insightful convolution relationship which serves to directly indicate the system performance and facilitate the use of frequency analysis.

Using various approximations, each having different degrees of validity, equation (4.41) can be linearized to provide a convolution form. One approach is the modeling of the solid object as an array of planar objects as given by

$$\mu_0(x, y, z) = \sum_i \tau_i(x, y)\delta(z - z_i) \qquad (4.42)$$

with the resultant detected intensity

$$I_d(x_d, y_d) = \frac{1}{4\pi d^2} \int\int s(x_s, y_s) \exp\left\{ -\sum_i \left[\tau_i\left(\frac{x_d - m_i x_s}{M_i}, \frac{y_d - m_i y_s}{M_i}\right) \right] \right\}dx_s dy_s$$

$$(4.43)$$

where $m_i = -(d - z_i)/z_i$ and $M_i = d/z_i$. If we make the assumption $\int \mu dz < 1$, namely that the attenuation through any path is relatively small, we can linearize the exponential where $\exp\left(-\int \mu dz\right) \simeq 1 - \int \mu dz$, giving

$$I_d(x_d, y_d) \simeq I_i - \sum_i \frac{1}{4\pi d^2 m_i^2} s\left(\frac{x_d}{m_i}, \frac{y_d}{m_i}\right) ** \tau_i\left(\frac{x_d}{M_i}, \frac{y_d}{M_i}\right) \qquad (4.44)$$

where $I_i = (1/4\pi d^2) \int\int s(x_s, y_s)dx_s dy_s$, the intensity in the absence of an object.

This formulation (4.44) provides a convolution relationship for all planes of a solid object. Unfortunately, the approximation used in the derivation, $\int \mu dz < 1$, is quite inaccurate except for very thin portions of the body. At diagnostic energy levels the attenuation coefficient of most soft tissue is about 0.2 cm^{-1}. Thus a typical 20-cm depth provides $\int \mu dz \simeq 4$, which makes the approximation unreasonable.

An alternative approach is to assume that most body tissue has an attenuation coefficient similar to that of water, so that the attenuation coefficient is decomposed as

$$\mu_0(x, y, z) = \mu_w(x, y, z) + \mu_\Delta(x, y, z) \qquad (4.45)$$

where μ_w is the attenuation coefficient of water and μ_Δ is the departure from that of water. We can now more legitimately assume that $\int \mu_\Delta dz < 1$. Those coefficients that do depart significantly from water, namely air and bone, are often associated with relatively short path lengths, so that the assumption can remain valid. Substituting (4.45) in (4.41) and using the assumption above, we obtain

$$I_d(x_d, y_d) \simeq \frac{1}{4\pi d^2} \int\int s(x_s, y_s) \exp\left[-\int \mu_w\left(\frac{x_d - mx_s}{M}, \frac{y_d - my_s}{M}, z\right)dz\right]$$

$$\times \left[1 - \int \mu_\Delta\left(\frac{x_d - mx_s}{M}, \frac{y_d - my_s}{M}, z\right)dz\right]dx_s dy_s. \qquad (4.46)$$

The first exponential in the integral represents the line integral of the object consisting of uniform tissue having attenuation coefficient μ_w. Because of the uniformity we can make the approximation

$$\mu_w\left(\frac{x_d - mx_s}{M}, \frac{y_d - my_s}{M}, z\right) \simeq \mu_w\left(\frac{x_d}{M}, \frac{y_d}{M}, z\right). \qquad (4.47)$$

This approximation is valid within the interior of the object. It fails, however, at the boundaries of the object, where the attenuation coefficient goes abruptly from μ_w to zero. Using this approximation, and restructuring the integration involving μ_Δ as a summation of planes, as in (4.42) to (4.44), we obtain

$$I_d(x_d, y_d) \simeq T_w\left[I_i - \sum_i \frac{1}{4\pi d^2 m_i^2} s\left(\frac{x_d}{m_i}, \frac{y_d}{m_i}\right) ** \tau_{\Delta i}\left(\frac{x_d}{M_i}, \frac{y_d}{M_i}\right)\right] \qquad (4.48)$$

where

$$T_w = \exp\left[-\int \mu_w\left(\frac{x_d}{M}, \frac{y_d}{M}, z\right)dz\right].$$

Equation (4.48) provides a convolution relationship with each plane using more valid approximations than that of (4.44). T_w represents the transmission through the object as if it were composed uniformly of water and the source was a point. It should be emphasized that, although equation (4.48) provides a reasonable basis for calculating the intensity due to a solid object, that is not the main reason for its presentation. The principal conclusion to be drawn from this development is that it is reasonable to study the response to an individual plane within a volume using the simplified convolutional relationship of equation (4.36).

Although (4.44) and (4.48) provide the elegance of the convolution relationship, for the general three-dimensional object, (4.31) or (4.41) must be used. For example, using the rectangular object of Fig. 4.5 with equation (4.41), we obtain

$$I_d(x_d, y_d) = \frac{1}{4\pi d^2} \int\int s(x_s, y_s) \exp\left[-\mu_a \int \text{rect}\left(\frac{y_d - my_s}{ML}\right) \text{rect}\left(\frac{z - z_0}{w}\right)dz\right]$$

$$\times dx_s dy_s. \qquad (4.49)$$

In this case the limits of integration are no longer symmetrical in y_d. The integral over the object is given by

$$\int \text{rect}\left(\frac{y_d - my_s}{ML}\right) \text{rect}\left(\frac{z - z_0}{w}\right)dz = \int_{\min\left\{\begin{array}{c} z_0 - w/2 \\ \max\left\{\begin{array}{c} \frac{dL/2 - y_s d}{y_d - y_s} \\ \frac{-dL/2 - y_s d}{y_d - y_s} \end{array}\right. \end{array}\right.}^{\min\left\{\begin{array}{c} z_0 + w/2 \\ \max\left\{\begin{array}{c} \frac{dL/2 - y_s d}{y_d - y_s} \\ \frac{-dL/2 - y_s d}{y_d - y_s} \end{array}\right. \end{array}\right.} dz. \qquad (4.50)$$

These limits define five zones of integration corresponding to the regions delineated by the intersection of the rays from the finite source with the four corners of the rectangle.

In general, in the study of radiographic imaging, the simplified convolutional relationships of a single plane are preferred since they provide profound insight into the performance-limiting factors. To review the development of the convolutional approach, we first showed that, for a planar object, if the relatively small obliquity factors are neglected, the resultant image is the convolution of the magnified transparency with a magnified source. For the solid object, we first modeled it as an array of planes and showed that the convolutional form would again apply if the total line integral of the attenuation coefficient is quite small so that the exponential could be linearized. Since this is not the usual case, we then modeled the solid object as a sum of a water coefficient, its dominant component, and a difference from this coefficient. Since the line integral of the difference component is quite small, we could linearize this portion of the expression and express each plane in convolutional form. As indicated, the motivation for this exercise was not primarily to establish an analytic procedure to deal with solid objects. It was, rather, a justification for analyzing systems by their response to a single planar object. We have shown that the simplified planar object analysis does indeed predict the performance of complex *volumetric* objects.

NONPARALLEL SOURCE DISTRIBUTIONS

In most x-ray tubes, as shown in Fig. 4.1, the source is not parallel to the plane of the detector. This results in a different point-spread function for each region of the detector plane. In general the source is a three-dimensional surface $s(x_s, y_s, z_s)$. Using the same type of derivation as that of the planar source, the generalized expression for the recorded intensity becomes

$$I_d(x_d, y_d) = \frac{1}{4\pi d^2} \int \int \int \frac{s(x_s, y_s, z_s)}{\left[1 + \left(\frac{r_{ds}}{d - z_s}\right)^2\right]^{3/2}} \exp\left[-\sqrt{1 + \left(\frac{r_{ds}}{d - z_s}\right)^2}\right.$$

$$\left. \times \int \mu_0\left(\frac{x_d - m'x_s}{M'}, \frac{y_d - m'y_s}{M'}, z\right) dz \right] dx_s dy_s dz_s \qquad (4.51)$$

where

$$m' = -\frac{d - z}{z - z_s} \quad \text{and} \quad M' = \frac{d - z_s}{z - z_s}.$$

As shown with the rotating anode tube of Fig. 4.1, the conventional x-ray tube has a source that can be approximated as a planar surface which is tilted with respect to the detector. The effects of this source distribution are illustrated in

Fig. 4.15. As can be seen, the projected source size varies significantly for different y positions on the planar object $t(x, y)$. The source geometry is modeled by setting $z_s = \alpha y_s$, where α is the tangent of the angle between the source plane and the vertical.

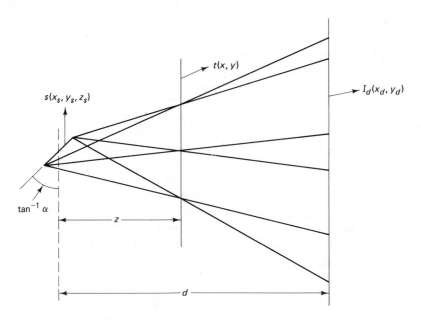

FIG. 4.15 Point response variations due to a tilted source.

For the tilted source case we can again evaluate the impulse response using a pinhole at x', y' as in Fig. 4.13. Since, in general, the magnifications will be different along each axis, we can rewrite equation (4.37) as

$$h(x_d, y_d, x', y') = \frac{\eta}{m_x m_y} s\left(\frac{x_d - M_x x'}{m_x}, \frac{y_d - M_y y'}{m_y}\right) \qquad (4.52)$$

where m_x, m_y, M_x, and M_y are the incremental magnifications as given by

$$m_x = \frac{\partial x_d}{\partial x_s} \quad \text{and} \quad m_y = \frac{\partial y_d}{\partial y_s}$$

$$M_x = \frac{\partial x_d}{\partial x'} \quad \text{and} \quad M_y = \frac{\partial y_d}{\partial y'}. \qquad (4.53)$$

In the case of the planar source parallel to the detector, these magnifications were constants independent of source and object coordinates. For the tilted source these constants are evaluated with the aid of Fig. 4.16. The values of m' and M' for the tilted source, by geometry, become

$$m' = -\frac{d - z}{z - \alpha y_s} \quad \text{and} \quad M' = \frac{d - \alpha y_s}{z - \alpha y_s}. \qquad (4.54)$$

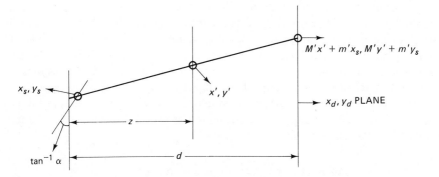

FIG. 4.16 Ray tracing for a tilted source.

The incremental magnifications are found by appropriately differentiating the recorder coordinates in Fig. 4.16, where $x_d = M'x' + m'x_s$ and $y_d = M'y' + m'y_s$. These are given by

$$m_x = m' = -\frac{d - z}{z - \alpha y_s}, \quad m_y = -\frac{(d - z)(z - \alpha y')}{(z - \alpha y_s)^2},$$

$$\text{and} \quad M_x = M_y = \frac{d - \alpha y_s}{z - \alpha y_s}. \tag{4.55}$$

For any sources of interest the source size will be significantly smaller than the object depth z. Thus $z \gg \alpha y_s$ and $d \gg \alpha y_s$. We then get the approximate relationships

$$m_x \simeq -\frac{d - z}{z} = m, \quad m_y \simeq -\frac{(d - z)(z - \alpha y')}{z^2} = m\left(\frac{1 - \alpha y'}{z}\right),$$

$$\text{and} \quad M_x = M_y \simeq \frac{d}{z} = M. \tag{4.56}$$

Using these relationships the point response, ignoring obliquity, is given by

$$h(x_d, y_d, x', y') = \frac{1}{4\pi z^2 m^2(1 - \alpha y'/z)} s\left[\frac{x_d - Mx'}{m}, \frac{y_d - My'}{m(1 - \alpha y'/z)}\right]. \tag{4.57}$$

This impulse response confirms the behavior shown in Fig. 4.15, where the y magnification changes significantly with the vertical position of the object point y'. The x magnification remains essentially unchanged. For the case of an impulse at $y' = z/\alpha$, the detector sees the edge of the source, resulting in a line image. Equation (4.57) then reduces to a delta function in the y dimension.

For the intensity due to a general transparency $t(x, y)$, we use the impulse response in the superposition integral. Making the substitutions $x'' = Mx'$ and $y'' = My'$, the detected intensity becomes

$$I_d(x_d, y_d) = \frac{1}{4\pi d^2 m^2} \int \int \frac{1}{1 - \alpha y''/Mz} s\left(\frac{x_d - x''}{m}, \frac{y_d - y''}{m(1 - \alpha y''/Mz)}\right)$$

$$\times t\left(\frac{x''}{M}, \frac{y''}{M}\right) dx'' dy''. \tag{4.58}$$

Despite the appropriate substitutions, equation (4.58) remains space variant because of the variation in the y magnification with the object coordinate. In an attempt to use the convolution formulation, we can divide the object plane into narrow horizontal strips at each value of y'. These strips form space-invariant or isoplanatic patches within which the impulse response is constant. Each horizontal strip at an object coordinate y' corresponds to a horizontal strip at the detector plane at coordinate $y_d = My'$. An approximate convolutional relationship can be structured at each horizontal strip by reformulating equation (4.58) as

$$I_d(x_d, y_d, y'_d) \simeq \frac{1}{4\pi d^2 m^2 (1 - \alpha y'_d/d)} s\left(\frac{x_d}{m}, \frac{y_d}{m(1 - \alpha y'_d/d)}\right) ** t\left(\frac{x_d}{M}, \frac{y_d}{M}\right) \quad (4.59)$$

where y'_d is the detector coordinate of the region of interest and is a constant in the convolution operation. The variation of vertical resolution with y'_d is clearly indicated. This relationship can be transformed into the frequency domain as

$$I_d(u, v, y'_d) = \frac{M^2}{4\pi d^2} S\left[mu, m\left(1 - \frac{\alpha y'_d}{d}\right)v\right] T(Mu, Mv). \quad (4.60)$$

Again at $y' = z/\alpha$ corresponding to $y'_d = d/\alpha$, we see the infinite bandwidth in the v dimension.

We have shown that the commonly used rotating anode tube can be structured as a tilted planar source. Although all the magnifications become a function of object position, the only one that changes significantly is the source magnification in the direction of the tilt. Using formulations for the incremental magnifications, with appropriate approximations, we develop an impulse response which is a function of the object position in the direction of the tilt. This allows an approximate convolution relationship which serves to illustrate the nature of the blur function.

EFFECTS OF OBJECT MOTION

The oversimplified solution to the problem of a finite source size is to use an extremely small source. These sources, however, have reduced power output, requiring longer exposure intervals, resulting in blurring due to the motion of the object under study. These motions are either those of uncooperative patients, such as children, or the physiological body motions of the respiratory, cardiovascular, and gastrointestinal systems.

The blurring effect of motion can be considered as a linear system parameter similar to that of the source size. Using the basic system of Fig. 4.13, consider the pinhole aperture moving uniformly in the x direction with a velocity v during the exposure interval T. The image movement at the recorder plane, by geometry, is MvT. The resultant recorded intensity is given by

$$I_d(x_d, y_d) = \frac{1}{4\pi d^2 m^2} t\left(\frac{x_d}{M}, \frac{y_d}{M}\right) ** s\left(\frac{x_d}{m}, \frac{y_d}{m}\right) * \frac{1}{MvT} \text{rect}\left(\frac{x_d}{MvT}\right). \qquad (4.61)$$

Thus the motion blurring is minimized by a short exposure time T.

Equation (4.61) can be restructured to emphasize the trade-off between source size and exposure time as limiting the system resolution. The energy density of the source $s(x, y)$ can be written as the product of a power density $p(x, y)$ and the exposure time T. The total source energy E_s is thus given by $T \iint p(x, y)dxdy$. To minimize the exposure time, for a given source energy, the source is operated at its maximum power density P_{max}, usually determined by the temperature limit. If we assume that the source is emitting uniformly and has an extent of $a(x, y)$, equation (4.61) becomes

$$I_d(x_d, y_d) = \frac{t(x_d/M, y_d/M)}{4\pi d^2 m^2} ** P_{max}Ta\left(\frac{x_d}{m}, \frac{y_d}{m}\right) * \frac{P_{max} \iint a(x, y)dxdy}{MvE_s}$$

$$\times \text{rect}\left[\frac{x_d P_{max} \iint a(x, y)dxdy}{MvE_s}\right] \qquad (4.62)$$

where $a(x, y)$ is a binary function defining the extent of the uniform source. The T in the rect function due to motion has been replaced by the source energy divided by the integrated source power. For a square source where $a(x, y) = \text{rect } x/L \text{ rect } y/L$, the impulse response is given by

$$h(x_d, y_d) = K \text{ rect}\left(\frac{x_d}{mL}, \frac{y_d}{mL}\right) * \text{rect}\left(\frac{x_d}{MvE_s/P_{max}L^2}\right). \qquad (4.63)$$

As indicated in (4.63), a larger source size results in a decreased motion blurring, and vice versa. The total extent of the point response in the x direction, X, due to source size and motion, is given by

$$X = |m|L + \frac{MvE_s}{P_{max}L^2}. \qquad (4.64)$$

This expression can be minimized with respect to L, giving

$$L_{min} = \left(\frac{2MvE_s}{P_{max}|m|}\right)^{1/3}. \qquad (4.65)$$

A square source size having this dimension will provide the smallest point response in the x direction. The corresponding exposure time T is given by

$$T = \frac{E_s}{P_{max}L_{min}^2} = \left(\frac{E_s}{P_{max}}\right)^{1/3}\left(\frac{|m|}{2Mv}\right)^{2/3}. \qquad (4.66)$$

Thus, as the velocity of the object increases, the optimum source size becomes larger and the exposure time correspondingly smaller.

REPRESENTATIVE SOURCE CONFIGURATIONS

Standard x-ray tubes use directly heated cathodes consisting of a coiled filament within a focusing cup. The resultant electron optics often results in two distinct areas on the rotating anode which are being bombarded with electrons and producing x-rays, as illustrated in Fig. 4.17. The source function can be approximated as two narrow rectangles each $w \times L$ separated by W, as given by

$$s(x, y) = \left[\text{rect} \left(\frac{x - W/2}{w} \right) + \text{rect} \left(\frac{x + W/2}{w} \right) \right] \text{rect} \left(\frac{y}{L} \right). \qquad (4.67)$$

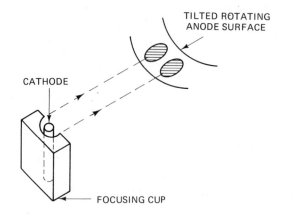

FIG. 4.17 Typical focal spot shape in x-ray tubes.

Typical dimensions for W and L vary from 0.3 to 2.5 mm. This represents a relatively poor response in the x direction, which can distort vertical edges. One indication of the problem is the Fourier transform of the source distribution, as given by

$$S(u, v) = 2wL \cos (\pi W u) \, \text{sinc} \, (wu) \, \text{sinc} \, (Lv). \qquad (4.68)$$

The cosine function makes the response highly oscillatory in the x direction. Of course, as previously indicated, the system response to a source function is determined by the source magnification m. Thus object planes close to the detector, where m is relatively small, will be relatively independent of the source size and shape.

Many efforts are under way to provide source configurations which are both smaller and have preferred shapes. In some x-ray tubes additional focusing fields are applied to cause the electron beam to produce a more desirable single spot. Microfocus tubes are available which use electron guns and produce focal spots of 50 to 200 microns. Field emission tubes have no heated filament and emit electrons from sharp points on a cylindrical cathode where the electric field is very high. These electrons impinge on a conical anode within the cylinder, resulting in a relatively large annular source size. These tubes use relatively high voltages and low currents and have not achieved widespread use.

Focal spots are generally measured using a pinhole camera. A small hole is placed in a relatively opaque sheet of high-atomic-number metal such as lead or gold. The pinhole is placed between the source and the film recorder. Ignoring obliquity factors, the resultant source image is given by

$$I_d(x_d, y_d) = \frac{1}{4\pi z^2 m^2 M^2} p\left(\frac{x_d}{M}, \frac{y_d}{M}\right) ** s\left(\frac{x_d}{m}, \frac{y_d}{m}\right) \tag{4.69}$$

where $p(x, y)$ represents the pinhole. The detected image will essentially represent the source as long as the magnified pinhole $p(x/M, y/M)$ is appreciably smaller in extent than the magnified source image $s(x/m, y/m)$. In this way the pinhole acts as a two-dimensional delta function, reproducing the source image. Typically, the pinhole is midway between source and detector with $|m| \simeq 1$ and $M \simeq 2$, so that the source image is approximately actual size. A relatively high resolution recorder, consisting of film only, is used to preserve the source image.

PROBLEMS

4.1 (a) Using a point-source x-ray system a distance d from the detector, find an approximate expression for the distance from the center of the detector r_d where the incident intensity has fallen off a fractional amount Δ, where $\Delta \ll 1$.

(b) Using the same system, a slab of material of thickness W and attenuation coefficient μ_0, parallel to the detector, is placed in the x-ray path. Neglecting the falloff in incident intensity, calculate the value of r_d at which the detected intensity has fallen off a fractional amount Δ, where $\Delta \ll 1$.

(c) For parts (a) and (b), calculate r_d for $d = 1$ meter, $W = 20$ cm, $\mu_0 = 0.25$ cm^{-1}, and $\Delta = 1\%$.

4.2 As shown in Fig. P4.2, a cylindrical bone of infinite length is embedded in a layer of soft tissue of infinite extent. The linear attenuation coefficient of the bone is μ_b and that of the soft tissue is μ_w. The incident intensity is I_0 using parallel x-rays.

(a) Find an expression for I_d, the detected intensity.

(b) What is the ratio of the detected intensity through the maximum bone thickness at $y_d = 0$ to that of the soft-tissue-only region where $y_d > R$?

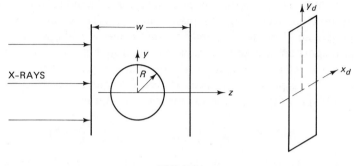

FIG. P4.2

(c) Calculate this ratio for $W = 20\,\text{cm}$, $R = 0.5\,\text{cm}$, using the curves of Fig. 3.7, where the soft tissue is muscle having a density of 1.0 and the bone density is 1.75. Perform the calculation for x-ray photon energies of 30 and 100 kev.

4.3 A cylindrical object having an attenuation coefficient μ_0 is positioned in a point-source x-ray system as shown in Fig. P4.3.

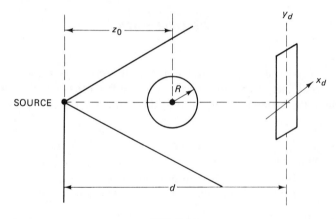

FIG. P4.3

(a) Find an expression for I_d, neglecting the falloff of the source intensity over the detector plane due to obliquity.

(b) Find an expression for I_d using the object in Problem 4.2 with the cylinder in a layer of soft tissue.

4.4 A cylinder of attenuation coefficient μ, radius R, and length L is placed on the axis of an x-ray imaging system as shown in Fig. P4.4. Neglecting all obliquity factors, find an expression for I_d versus r_d, where the intensity in the absence of the object is a uniform I_0.

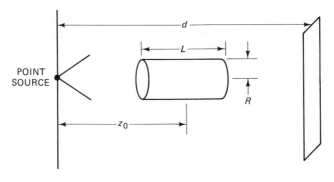

FIG. P4.4

4.5 A rectangular x-ray source, $s(x_s, y_s) = \text{rect}(x/X)\,\text{rect}(y/Y)$, is used with two opaque, semi-infinite planes as shown in Fig. P4.5. Ignoring all obliquity factors, plot the intensity versus y_d on the detector plane labeling all break points.

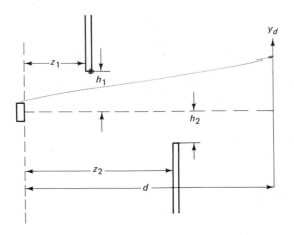

FIG. P4.5

4.6 An $L \times L$ x-ray source, having unity intensity, parallel to and a distance d from the recorder, is used to image a planar transparency a distance z from the source having a transmission

$$t = \tfrac{1}{2} + \tfrac{1}{2}\cos 2\pi a y$$

(a) Ignoring obliquity, find an expression for the intensity at the recorder plane. (Do not leave in convolutional form.)

(b) Repeat part (a) where the source is tilted at angle $\tan^{-1}\alpha$, where $\alpha = z_s/y_s$, and the projected size of the source in the xy plane continues to be $L \times L$.

4.7 A tilted source is used to image an opaque planar tilted object infinite in extent and containing three pinholes of equal size as shown in Fig. P4.7. Neglecting obliquity, plot I_d versus y_d, indicating the relative amplitudes and the position of the break points. Space invariance can be assumed in the vicinity of the pinhole images.

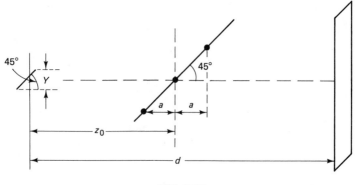

FIG. P4.7

4.8 A rectangular source tilted by an angle θ is used to image an opaque object tilted at 45° as shown in Fig. P4.8. The projection of the source Y is significantly smaller than all other dimensions. Neglecting obliquity factors, plot the relative detected intensity in the y_d direction labeling the y_d axis at the break points. [*Hint*: Assume space invariance in the vicinity of the break points.]

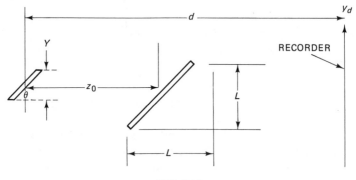

FIG. P4.8

4.9 A source tilted at an angle of 45° has a projected intensity $s(x, y) = K$ circ (r/r_0). It is used to image a transparency at $z = z_0$, having a transmission $t(x, y) = \sum_i \delta(x)\delta(y - i)$. Find the resultant intensity at $z = d$, neglecting obliquity.

5

Recorder Resolution
Considerations

We have thus far considered resolution limitations due to the x-ray source. The other important resolution-limiting factor in the system is the x-ray recorder, where the image itself is formed. The principal difficulty is in the attaining of the desired high resolution while maintaining a relatively high quantum or capture efficiency. The quantum or capture efficiency represents the fraction of photons that interact within the recorder material. As will be shown in Chapter 6, the number of captured photons per picture element governs the resultant signal-to-noise ratio. This SNR will be shown to be given by

$$\text{SNR} = C\sqrt{\eta N} \qquad (5.1)$$

where N is the number of photons per picture element impinging on the recorder, η is the quantum or capture efficiency, and C is the contrast of the structure of interest. A thick recorder has a high quantum efficiency but, as will be shown, exhibits poor resolution. Similarly, very thin recorders exhibit negligible blurring due to spreading, but capture relatively few of the photons.

SCREEN–FILM SYSTEMS

X-ray film itself is a relatively inefficient recorder of x-ray photons. To collect the x-ray photons efficiently, a scintillating screen is used to convert each x-ray photon into a large number of visible photons, which are then recorded on film. The scintillating screen is a dense high-atomic-number material, such as calcium tungstate, which will capture the x-ray photons in a relatively short path for resolution considerations. This is illustrated in Fig. 5.1. An incoming photon gives up its energy to a scintillating phosphor at a distance x as shown. A large number of visible photons are generated in the scintillating crystal. The exact nature of this process is quite complex [Bates and Morwood, 1973]. We will assume an isotropic radiator of visible light at the scintillation point. Also, we neglect the granular nature of the phosphor and assume uniform light propagation. It is reasonably accurate to assume that the x-ray photon does not interact with the film itself.

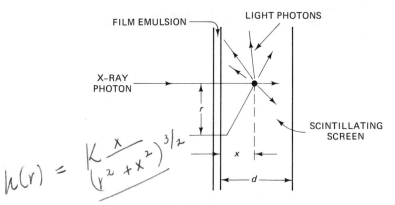

FIG. 5.1 X-ray recording process.

The invariant impulse response $h(r)$ along the recording film emulsion is governed by two factors, the obliquity factor of the impinging light photons, which follows a cosine law, and the inverse-square-law falloff with distance from the scintillation point, which has a $\cos^2 \theta$ dependence. Thus $h(r)$, as with the x-ray source in Fig. 4.3, is given by

$$h(r) = h(0) \cos^3 \theta = h(0) \frac{x^3}{(x^2 + r^2)^{3/2}} \qquad (5.2)$$

where $h(0)$ is the response at $r = 0$ given by $h(0) = K/x^2$. Thus $h(r)$ is given by

$$h(r) = K \frac{x}{(r^2 + x^2)^{3/2}} \qquad (5.3)$$

where K is an intensity proportionality constant. With the assumptions made above, this represents the point-spread function of the x-ray recording process.

Since the system is space invariant, the frequency response of the system is given by the Fourier transform of the point-spread function, as given by

$$H_1(\rho) = \mathfrak{F}\{h(r)\} = 2\pi \int_0^\infty \frac{Kx}{(r^2 + x^2)^{3/2}} J_0(2\pi\rho r) r\, dr \tag{5.4}$$

where $J_0(2\pi\rho r)r$ is the kernel of the Fourier–Bessel transform for functions having circular symmetry and ρ is the radial spatial frequency variable as given in equation (2.35). The resultant transform is given by

$$H_1(\rho) = 2\pi K e^{-2\pi x\rho}. \tag{5.5}$$

It is convenient to use a normalized frequency response $H(\rho)$ as given by

$$H(\rho) = \frac{H_1(\rho)}{H_1(0)} = e^{-2\pi x\rho}. \tag{5.6}$$

This normalization can be performed since it only involves the elimination of constant terms and no functions of x. In general, however, the normalization should be performed after the averaging process, where $\bar{H}(\rho)$ is obtained, to ensure that each individual response $H_1(\rho, x)$ is properly weighted in the averaging process.

The normalized frequency response, $H(\rho, x)$, is the transfer function resulting from a photon giving up its energy at a distance x. In order to find the average transfer function $\bar{H}(\rho)$ resulting from a large number of x-ray photons, we integrate over the probability density function $p(x)$ as given by

$$\bar{H}(\rho) = \int H(\rho, x) p(x) dx = \int e^{-2\pi x\rho} p(x) dx \tag{5.7}$$

where $p(x)$ is the density function of the point at which the x-ray photons interact. The probability density function can be determined from the distribution function $F(x)$ given in equation (2.45) which is the probability that an x-ray photon will interact within a distance x. For an infinitely thick scintillating phosphor, the distribution function is given by

$$F(x) = 1 - e^{-\mu x}. \tag{5.8}$$

where μ is the linear attenuation coefficient and $e^{-\mu x}$ represents the fraction of photons transmitted beyond distance x. The associated probability density is given by

$$p(x) = \frac{d}{dx} F(x) = \mu e^{-\mu x}. \tag{5.9}$$

For a phosphor screen of thickness d, the distribution function is given by

$$F(x) = \frac{1 - e^{-\mu x}}{1 - e^{-\mu d}}. \tag{5.10}$$

This distribution function represents *captured photons only* and ignores those transmitted beyond $x = d$ since they do not contribute to the resultant image. Thus the distribution function varies from zero to one as x varies from zero to d. The resultant probability density function is $\mu e^{-\mu x}/(1 - e^{-\mu d})$ with the normalized spectrum given by

$$\bar{H}(\rho) = \frac{\mu}{1 - e^{-\mu d}} \int_0^d e^{-2\pi x\rho} e^{-\mu x} dx$$

$$= \frac{\mu}{(2\pi\rho + \mu)(1 - e^{-\mu d})}[1 - e^{-d(2\pi\rho+\mu)}]. \qquad (5.11)$$

Equation (5.11) represents a monotonically decreasing response with increasing spatial frequency. A representative value for the thickness d of a typical screen is 0.25 mm. A calcium tungstate screen at the center of the diagnostic photon energy spectrum will have an attenuation coefficient μ of about 15 cm^{-1}. A plot using these values is shown in Fig. 5.2.

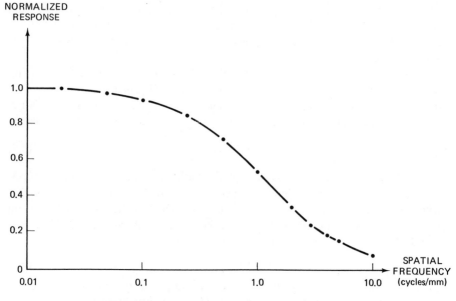

FIG. 5.2 Frequency response of a screen-film system.

It is important to establish some figure of merit for the frequency response, such as a cutoff frequency or effective bandwidth, to evaluate various configurations. Referring to equation (5.11), we note that, using typical values, the bracketed expression can be approximated as unity above relatively low spatial frequencies. For example, at $\rho = 1.0$ cycle/mm, $\bar{H}(\rho) = 0.53$ and the bracketed expression is 0.85. Clearly, in establishing a cutoff frequency, it is convenient to assume that the bracketed expression is unity. We define the cutoff frequency ρ_k as that spatial frequency where $\bar{H}(\rho_k) = k$. For values of k less than 0.5, this can be clearly approximated as

$$\bar{H}(\rho_k) = k \simeq \frac{\mu}{(2\pi\rho_k + \mu)(1 - e^{-\mu d})}. \tag{5.12}$$

Substituting $\eta = 1 - e^{-\mu d}$, the capture efficiency of the screen, and using the reasonable approximation $\eta k \ll 1$, we get an expression for the cutoff frequency:

$$\rho_k \simeq \frac{\mu}{2\pi k\eta}. \tag{5.13}$$

The currently used definition of limiting resolution is the 10% response. Therefore, for this case the limiting resolution in cycles/mm would be $\mu/0.2\pi\eta$. This clearly shows the trade-off between high-frequency response and the efficiency as regards the thickness d. A high attenuation coefficient μ is always desirable since it provides for the stopping of photons in a short distance. The thickness d, however, must be a trade-off between efficiency and the cutoff frequency. In practice a variety of screens are made available of differing thickness, so that the desired trade-off between efficiency and resolution can be achieved for each study.

Modern x-ray recording systems utilize a double screen–film cassette structure which helps this trade-off among screen thickness, resolution, and efficiency, as shown in Fig. 5.3. The recording film has a photographic emul-

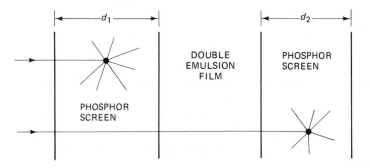

FIG. 5.3 Dual-screen recording system.

sion on each side and is sandwiched between two phosphor layers. The efficiency is now based on the entire phosphor thickness $d = d_1 + d_2$. Scintillations in either layer will be recorded on one of the two emulsions. The frequency response due to a scintillation at x is given by

$$H(\rho, x) = e^{-2\pi\rho(d_1-x)} \quad \text{for } 0 < x < d_1$$
$$= e^{-2\pi\rho(x-d_1)} \quad \text{for } d_1 < x < d. \tag{5.14}$$

The film itself is essentially transparent to x-rays.

The frequency response averaged over many events using equation (5.7) becomes

$$\bar{H}(\rho) = \frac{\mu}{1 - e^{-\mu d}} \left\{ \int_0^{d_1} \exp\left[-2\pi\rho(d_1 - x) + \mu x\right] dx \right.$$
$$\left. + \int_{d_1}^{d} \exp\left[-2\pi\rho(x - d_1) + \mu x\right] dx \right\}$$
$$= \frac{\mu}{1 - e^{-\mu d}} \left[\frac{e^{-\mu d_1} - e^{-2\pi\rho d_1}}{2\pi\rho - \mu} + \frac{e^{-\mu d_1} - e^{-(2\pi\rho d_2 + \mu d)}}{2\pi\rho + \mu} \right]. \tag{5.15}$$

We will establish an effective cutoff frequency ρ_k for this configuration as was done for the previous case. Since d_1 and d_2 are comparable in width, each about half of d, we can again neglect the exponents of the form $\exp(-2\pi\rho d)$, since they will become negligible at relatively low spatial frequencies. We also make the approximation

$$\frac{1}{2\pi\rho - \mu} + \frac{1}{2\pi\rho + \mu} \simeq \frac{2}{2\pi\rho} \tag{5.16}$$

which is valid as long as $(2\pi\rho)^2 \gg \mu^2$. As before, this is appropriate at all but the lower spatial frequencies. Using these approximations we again define the cutoff frequency ρ_k as that frequency at which $\bar{H} = k$ as given by

$$\bar{H}(\rho_k) = k \simeq \frac{\mu}{\eta} e^{-\mu d_1} \frac{2}{2\pi\rho_k} \tag{5.17}$$

resulting in a cutoff frequency

$$\rho_k \simeq \frac{\mu}{2\pi k\eta} 2e^{-\mu d_1}. \tag{5.18}$$

This indicates that this configuration, for a given efficiency, has an improvement given by $2e^{-\mu d_1}$. It is erroneous to assume that a reduction of d_1 to zero will maximize the resolution since this negates the approximation of $\bar{H}(\rho)$ at higher frequencies and simply returns us to the original single-layer configuration. The optimum condition, as would be expected, is approximately at $d_1 = d/2$. Under these conditions we have

$$\rho_k \simeq \frac{\mu}{2\pi k\eta} 2\sqrt{1 - \eta} \tag{5.19}$$

with a resultant improvement over the single screen of $2\sqrt{1 - \eta}$. The double-screen improvement factor at high spatial frequencies for commercial screens where $\eta \simeq 0.3$ is about 1.7. This substantial factor can be used to provide improved frequency response at a given efficiency, improved efficiency at a given resolution, or any intermediate combination.

The optimum division of the two screens between d_1 and d_2 for a given d is a somewhat complex subject. A simplified approach is to place the division at the mean stopping distance of the photons, thus ensuring that the scintillations in both screens will be as close as possible to the film emulsions. The probability density function of the photon interaction point, as previously derived, is given by

$$p(x) = \frac{\mu}{1 - e^{-\mu d}} e^{-\mu x}. \qquad (5.20)$$

The mean interaction distance $\bar{x}$ is given by

$$\bar{x} = \int_0^d x p(x) dx \qquad (5.21a)$$

$$\bar{x} = \frac{1 - e^{-\mu d}(\mu d + 1)}{\mu(1 - e^{-\mu d})}. \qquad (5.21b)$$

For the typical values previously used, $d = 0.25$ mm and $\mu = 15$ cm^{-1}, the optimum is 0.117 mm, approximately equal to $d/2$. For thicker screens used for maximum sensitivity, with reduced resolution, the front screen depth d_1 can be half of the back screen depth d_2.

CRITICAL-ANGLE CONSIDERATIONS

A further refinement of the frequency response of the phosphor screen makes use of the problem of the critical angle. Assuming perfect contact between the phosphor and the film, the maximum or critical angle θ_c at which light from the phosphor will enter the film is given by

$$\sin \theta_c = \frac{n_f}{n_p} \qquad (5.22)$$

where n_f is the refractive index of the film emulsion, and n_p is the refractive index of the phosphor material where $n_p > n_f$. If we call R the maximum radius from the scintillation point at which light enters the film emulsion, we have

$$\frac{n_f}{n_p} = \frac{R}{\sqrt{R^2 + x^2}}$$

giving

$$R = \frac{x}{\sqrt{(n_p/n_f)^2 - 1}}. \qquad (5.23)$$

Rather than the unlimited spreading of the light as shown in Fig. 5.1, the spreading becomes limited to a circle of radius R. Thus the original point response of the system is multiplied by circ (r/R) as given by

$$h(r) = K \frac{x}{(r^2 + x^2)^{3/2}} \, \text{circ} \left(\frac{r\sqrt{n^2 - 1}}{x} \right) \qquad (5.24)$$

where $n = n_p/n_f$ and the circ function is defined in Table 2.2. The Fourier transform of the circ function is given by

$$\mathcal{F} \left\{ \text{circ} \frac{r\sqrt{n^2 - 1}}{x} \right\} = \frac{x}{\rho\sqrt{n^2 - 1}} J_1 \left(\frac{2\pi x \rho}{\sqrt{n^2 - 1}} \right). \qquad (5.25)$$

The total frequency response is the convolution of the originally derived response and that due to the circ function as given by

$$H_1(\rho) = Ke^{-2\pi x\rho} * \frac{xJ_1(2\pi x\rho/\sqrt{n^2 - 1})}{\sqrt{n^2 - 1}\rho}. \tag{5.26}$$

In general, the critical-angle consideration increases the resolution but decreases the number of captured light photons per x-ray photon. The effect on noise is considered in Chapter 6.

ENERGY SPECTRUM CONSIDERATIONS

The analysis of the recorder has been done under the assumption of a monochromatic energy spectrum. With the usual broad energy spectrum generated by x-ray tubes, energy dependence $\mu(\varepsilon)$ of the scintillating phosphor should be taken into account for a more complete analysis. Thus a more accurate expression for the averaged frequency response in the system of Fig. 5.1 is given by

$$\bar{H}_1(\rho) = K \int_0^d \int_{\varepsilon_1}^{\varepsilon_2} e^{-2\pi x\rho} \frac{e^{-\mu(\varepsilon)x}\mu(\varepsilon)}{1 - e^{-\mu(\varepsilon)d}} S'(\varepsilon)d\varepsilon dx \tag{5.27}$$

where K is a normalizing constant and $S'(\varepsilon)$ is the energy spectrum leaving the body and entering the recorder. For a parallel x-ray geometry, or for regions close to the axis where the obliquity of the rays can be ignored, this spectrum is given by

$$S'(\varepsilon) = S(\varepsilon) \exp\left[-\int \mu_1(\varepsilon, z)dz\right] \tag{5.28}$$

where $S(\varepsilon)$ is the source spectrum and μ_1 is the attenuation coefficient of the body as a function of energy and depth.

ALTERNATIVE APPROACHES TO RECORDER SYSTEMS

In recorders using scintillating phosphors, the light photons were scattered isotropically at the point of scintillation. This gave rise to the fundamental trade-off between resolution and efficiency. Three configurations can be used to avoid this trade-off.

First, the scintillating phosphor can be structured in the form of optical fibers where the inner material is the phosphor itself and, as with conventional fiber optics, the phosphor is clad with a layer of lower-refractive-index material. When an x-ray photon gives up its energy in one of the fibers, the resulting light is trapped within the fiber by complete internal reflection. Thus the trapped light bounces back and forth until it arrives at the film. The resolution is thus determined by the diameter of the fibers. Essentially, the fiber structure avoids the isotropic spreading of the light photons, with its resultant loss of resolution. To increase the collection efficiency, the fiber bundles are merely made longer, without a resultant loss in resolution. The photographic film can be placed on one side of the fiber bundle with an optically reflecting surface on the other side to ensure that most of the light photons eventually reach the film. Initial experiments have been made with structures of this type, although thus far the practical construction difficulties have prevented its commercial availability.

The second alternative approach uses a high-pressure gas chamber, usually xenon, as the detector. The x-ray photon gives up its energy in an ionization process that creates electron–hole pairs. A strong electric field is placed across the gas chamber so that the charged particles, once generated, will follow field lines and not disperse. The field is thus similar to the fiber structure, which forces the ionized particles to follow straight lines. Thus a relatively thick gas chamber can be used to ensure that most of the x-ray photons interact and thus provide high collection efficiency. The resultant charged particles follow the electric field and are deposited on a dielectric sheet. The charge pattern on this sheet represents the latent image. If developed with a toner, as is done in xerography, the desired image results. One of the practical difficulties is the removal of the dielectric sheet from the high-pressure gas chamber.

A third alternative approach is a scanning system where the information is developed in time rather than in space. A narrow pencil x-ray beam scans the subject. The detector is made very thick so as to have a very high capture efficiency. The detected signal is used to intensity modulate a synchronously scanned beam of a cathode ray tube to create the desired image. The important characteristic of this system is that the geometry of the detector does not affect the system's resolution. Thus a thick, high-efficiency scintillating crystal, such as sodium iodide, can be used followed by a photomultiplier to develop the electrical signal. The noise considerations of configurations of this type are considered in Chapter 6. In any case the resolution is governed solely by the geometry of the scanning beam and is independent of the detection process. Thus far systems of this type have been impractical since they use a very small portion of the x-rays emitted from the source. This causes a long exposure interval with the associated distortions due to respiratory and cardiovascular motion.

In recent years line detector arrays have been used which have been developed for computerized tomography systems. Here the x-rays are initially collimated into a planar or fan-shaped beam. This planar beam is projected through the body onto the line detector array. The resolution normal to the beam is determined solely by the beam thickness. The resolution along the beam is governed by the array of detectors. Each detector is usually shielded to prevent ionization products, be they light or charged particles, from entering the neighboring elements. Thus, as with optical fibers, the resolution along the beam is determined by the size of the detectors. Detector arrays with 0.5-mm detector elements have been constructed. Although the resolution is poorer than film screen systems, these arrays exhibit high efficiency and have electrical outputs that can be coupled to digital processing systems. The planar beam and line array are scanned, relative to the subject, to create a two-dimensional image.

In systems using discrete detector arrays with electrical outputs the resolution is governed by the detector size itself. Here the limitation in resolution is based on fabrication considerations. A more fundamental limitation on detector size is the signal-to-noise ratio, which is considered in Chapter 6.

OVERALL SYSTEM RESPONSE

The overall response of a projection imaging system, including the source and recorder considerations, is given by

$$I_d(x_d, y_d) = Kt\left(\frac{x_d}{M}, \frac{y_d}{M}\right) ** \frac{1}{m^2}s\left(\frac{r_d}{m}\right) ** h(r_d) \tag{5.29}$$

where, for convenience, the source response, like that of the recorder, is assumed to be radially symmetric. In this case the intensity I_d represents the recorded light photons. In the frequency domain this becomes

$$I_d(u, v) = KM^2T(Mu, Mv)H_0(u, v) \tag{5.30}$$

where $H_0(u, v)$, the overall transfer function between the magnified planar object and the image, is given by

$$H_0(u, v) = H_0(\rho) = S(m\rho)H(\rho). \tag{5.31}$$

This transfer function is based on spatial frequencies at the recorder, representing spatial frequencies of the magnified object $t(x_d/M, y_d/M)$. It is often desirable to evaluate the ability of the system to resolve specific spatial frequencies of the object itself. The output spectrum in terms of object spatial frequencies is given by

$$I_d\left(\frac{u}{M}, \frac{v}{M}\right) = KM^2T(u, v)H_0\left(\frac{\rho}{M}\right) \tag{5.32}$$

where

$$H_0\left(\frac{\rho}{M}\right) = S\left(\frac{m}{M}\rho\right)H\left(\frac{\rho}{M}\right). \tag{5.33}$$

This transfer function, in terms of object spatial frequencies ρ/M, enables us to study the conditions for maximizing the response to parameters of the object. For example, placing the response in terms of depth z, we obtain

$$H_0\left(\frac{\rho}{M}\right) = S\left[\left(1 - \frac{z}{d}\right)\rho\right]H\left(\frac{z}{d}\rho\right). \tag{5.34}$$

The maximization of this function determines the depth z which provides the best frequency response. With a small source, and a correspondingly broad $S(\rho)$, the optimum will occur at relatively small values of z since $H(\rho)$ will be the dominant factor. In the opposite extreme, for a large source having a narrow $S(\rho)$, the optimum will occur where the object plane is near the recorder with z approaching d. The optimum depth plane, in general, is determined by differentiating $H_0(\rho/M, z)$ with respect to the depth z.

PROBLEMS

5.1 Using the appropriate approximations find the ratio of the high-frequency cutoff ρ_k for the case shown in Fig. 5.1, with the film in front where the x-ray photons impinge, to that of a screen with the same geometry having the film in the back.

5.2 A film emulsion is placed on both sides of a phosphor screen of thickness d. The resultant transparencies are combined such as to provide an overall transparency having a small-signal normalized frequency response

$$\bar{H}_{\text{overall}} = \frac{\bar{H}_{\text{front}} + \bar{H}_{\text{back}}}{2}$$

where $\bar{H}_{\text{front}}$ and $\bar{H}_{\text{back}}$ are the individual normalized responses. Find the high-frequency cutoff ρ_k of the overall transparency using the appropriate approximations. By what factor does this cutoff frequency differ from the case of a single emulsion on the front side where the x-rays impinge?

5.3 In an effort to capture the escaping half of the light photons which are normally lost, a mirror of reflectivity R is placed in back of the screen as shown in Fig. P5.3. For this case calculate the following:

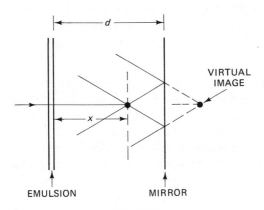

FIG. P5.3

(a) $H(\rho, x)$, the normalized frequency response for a photon stopping at distance x.

(b) $\bar{H}(\rho)$, the average frequency response.

(c) ρ_k, the high-frequency cutoff using the appropriate approximations.
[*Hint*: For the reflected light use the virtual image position of the scintillation, as shown.]

5.4 A more complete model of a film–screen system includes the effect of light from each screen reaching the opposite emulsion as shown in Fig. P5.4. Assuming that the film is transparent to light and x-rays, calculate $\bar{H}_1(\rho)$ and $\bar{H}_2(\rho)$, the normalized frequency responses of each emulsion. Calculate the overall response $\bar{H}_0(\rho)$ assuming that it is the average of the individual responses. By what factor does $\bar{H}_0(\rho)$ differ from the previous case, where the film thickness is ignored?

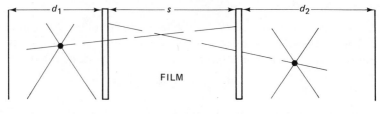

FIG. P5.4

5.5 An x-ray imaging system has a uniform circular source of radius r_1 and the impulse response of the recorder is a uniform circle of radius r_2. An opaque planar object at plane z has two pinholes separated by a distance s.

(a) What is the overall impulse response of the system to a planar object at depth z?

(b) What is the minimum distance s at which the images of the holes are separable, that is, the resultant responses do not overlap?

(c) At what ratio of r_1/r_2 is this minimum distance independent of the depth z?

5.6 An extended source x-ray system is used with the source parallel to the recorder plane and having a distribution $s(r) = e^{-ar^2}$. The recording plane, a distance d from the source, has an impulse response $h(r) = e^{-br^2}$. Neglecting all obliquity factors, at what distance z_0 from the source should a transparency be placed so as to maximize the relative response at a spatial frequency ρ_0 in the transparency? Discuss the optimum z_0 where $a \gg b$ and $b \gg a$.

6

Noise Considerations in

Radiography and Fluoroscopy

The ability to visualize a structure in a noise-free environment depends, among other factors, on the local contrast C, which we define as

$$C = \frac{\Delta I}{\bar{I}} \tag{6.1}$$

where $\bar{I}$ is the average background intensity and ΔI is the intensity variation in the region of interest. Contrast, however, is not a fundamental limit on visualization since it can be artificially enhanced by, for example, subtracting part of the background or raising the intensity pattern to some power. Noise, however, represents a fundamental limitation on the ability to visualize structures. The signal-to-noise ratio, a basic measure of visualization, is determined by the ratio of the desired intensity variations to the random intensity variations, which are governed by the statistical properties of the system. In general, our signal-to-noise ratio will be defined as

$$SNR = \frac{\Delta I}{\sigma_I} = \frac{C\bar{I}}{\sigma_I} \tag{6.2}$$

where σ_I is the standard deviation of the background intensity representing the rms value of the intensity fluctuations.

The noise properties of most communications systems involve additive noise only. The energy per photon, $h\nu$, in these lower-frequency spectra is relatively small, so that copius amounts of photons are available for the weakest signals being considered. Thus the uncertainties lie almost completely in the noise added by the measurement system rather than those of the signal itself. Quantum noise, which is the primary noise source in x-ray systems, is the noise due to the quantization of the energy into photons each having an energy $h\nu$. This quantum noise is Poisson distributed.

We can appreciate the dominance of the quantum noise by considering a communications system having an additive noise power of $4KTB$, where K is Boltzmann's constant, T is the absolute temperature, and B the bandwidth. The noise energy per time element $1/B$ is therefore $4KT$. The ratio of signal-to-noise energy per element can be structured as

$$\text{SNR} = \frac{N h\nu}{\sqrt{N h\nu + 4KT}} \tag{6.3}$$

where N is the number of photons per time element, $N h\nu$ is the signal energy per element, and $\sqrt{N h\nu}$ is the standard deviation or noise energy per element. In most communications systems $4KT \gg h\nu$, so that, for any reasonable signal strength, the additive thermal noise term in the denominator dominates. For example, at $\nu = 10^6$ Hz, $4KT/h\nu = 2.5 \times 10^7$. Even at $\nu = 10^2$ GHz, this ratio is 2.5×10^2. At the nominal x-ray frequencies of $\nu = 10^{19}$ Hz, corresponding to $\lambda = 0.2$ Å, this ratio is 2.5×10^{-6}. Clearly, equation (6.3) reduces to an SNR given by $\sqrt{N}$. This means that in the x-ray region the total energy represents a countable number of photons whose statistical uncertainty is the major contribution to the noise.

The emission of photons from the x-ray source is a Poisson process [Parzen, 1960] whose probability density was given in equation (2.53) as

$$P_k = \frac{N_0^k e^{-N_0}}{k!}$$

where P_k is the probability, in a given time interval, of emitting k photons, and N_0 is the average number of photons emitted during that interval.

In Chapter 3 we characterized the transmission of photons through the body as a binary process where photons either interacted and were removed from the beam, or did not interact and were transmitted to the recorder. This represents a binomial process [Parzen, 1960] where the probability p of a photon being transmitted is $\exp\left(-\int \mu dz\right)$ and the probability q of it being stopped is $1 - \exp\left(-\int \mu dz\right)$. The cascading of a Poisson and binomial distribution results in a Poisson distribution as shown below.

The probability of getting k photons through an object, $Q(k)$, is the sum of the probabilities of the various combinations transmitting k photons as given by

$$Q(k) = P(k)\binom{k}{k}p^k + P(k+1)\binom{k+1}{k}p^kq + \cdots + P(k+n)\binom{k+n}{k}p^kq^n \tag{6.4}$$

where each term represents a combination of a photon source producing $k + x$ photons, $P(k+x)$, and a binomial transmittance $\binom{k+x}{k}p^kq^x$, which combine to transmit k photons. Since P is a Poisson process, we have

$$P(k+n)\binom{k+n}{k}p^kq^n = \frac{N_0^{k+n}e^{-N_0}}{(k+n)!}\frac{(k+n)!\,p^kq^n}{k!\,n!}$$

$$= \frac{e^{-N_0}(pN_0)^k}{k!}\frac{(qN_0)^n}{n!}. \tag{6.5}$$

Substituting this result in the general equation for $Q(k)$, we obtain

$$Q(k) = \frac{e^{-N_0}(pN_0)^k}{k!}\sum_{n=0}^{\infty}\frac{(qN_0)^n}{n!} \tag{6.6}$$

where the summation is equal to e^{qN_0}. Substituting this into the formula for $Q(k)$, we obtain

$$Q(k) = \frac{e^{-pN_0}(pN_0)^k}{k!} \tag{6.7}$$

which is identically a Poisson process of rate pN_0. We have shown that the emerging photons from the object continue to be Poisson distributed with the rate scaled by the attenuation of the object $p = \exp(-\int \mu dz)$. The photons emitted from the object have a mean value N given by

$$N = N_0 \exp\left(-\int \mu dz\right). \tag{6.8}$$

In a Poisson process of mean N the variance is N and the standard deviation is $\sqrt{N}$ [Parzen, 1960]. We can thus calculate the signal-to-noise ratio of a structure having a contrast of C as given by

$$\text{SNR} = \frac{\Delta N}{\sqrt{N}} = C\sqrt{N} \qquad \text{contrast} \tag{6.9}$$

where the signal is ΔN, the variation in the number of photons per element defining the structure of interest, and the noise is $\sqrt{N}$, the standard deviation of the number of photons per element. As is seen for a given subject contrast C, the signal-to-noise ratio is proportional to the square root of the number of photons. Thus the signal-to-noise ratio can be made arbitrarily high except for the radiation limitation. We will briefly consider the nature of the radiation dose.

The subject of radiation [Johns and Cunningham, 1974; Sprawls, 1977] is relatively complex and will be treated here in a relatively simplified fashion.

Our principal intent is to relate dose considerations to the incident photon density and hence to the signal-to-noise ratio. Two quantities are of interest: the exposure, or amount of radiation delivered to a point, and the absorbed dose or the radiation energy absorbed in a region.

The unit of exposure is the roentgen (R), which is defined as producing ionization of 2.58×10^{-4} coulomb/kilogram in air. The absorbed dose unit is the rad, which is defined as an expenditure of energy of 100 ergs/gram. In air an exposure of 1 R corresponds to an absorbed dose of 0.87 rad. The number of rads per roentgen varies with different materials and with the energy used as shown in Fig. 6.1. As is seen, for soft tissue, there is approximately 1 rad per

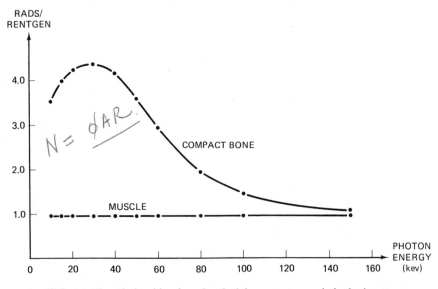

FIG. 6.1 The relationship of an absorbed dose to exposure in body tissues over the diagnostic energy range.

roentgen throughout the diagnostic energy range. Thus with the exception of bony regions, where the absorbed dose increases at lower energies, the rad and roentgen become essentially equivalent for our purposes.

Figure 6.2 shows the relationship of photon density to the exposure in roentgens. At low photon energies most of the photons interact, but the energy imparted per interaction is small. At very high photon energies, very few photons interact. These conflicting factors result in a peak at about 60 kev, with the average over the diagnostic range being about $\phi = 2.5 \times 10^{10}$ photons/cm² per roentgen. We use this relationship to characterize the number of photons per pixel N in terms of the incoming radiation as given by

$$N = \Phi A R \exp\left(-\int \mu dz\right) \tag{6.10}$$

X 10^{10} PHOTONS/cm^2/ROENTGEN

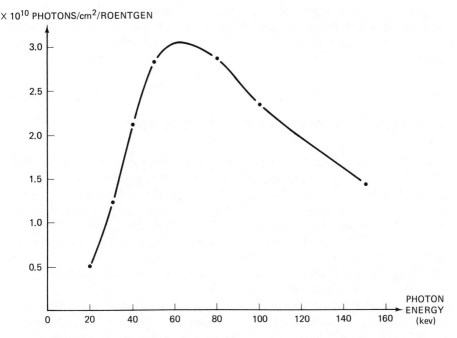

FIG. 6.2 The relationship of photon fluence to exposure over the diagnostic energy range.

where R is the number of roentgens incident on the subject, A is the area of a picture element in cm^2, $\exp\left(-\int \mu dz\right)$ is the transmission t through the body, and Φ is the photon density per roentgen. The signal-to-noise ratio in terms of radiation is given as

$$\text{SNR} = C\sqrt{\Phi A R t}. \tag{6.11}$$

The signal-to-noise ratio given above is that of the photons emerging from the body and thus would represent the performance of a detector which captured all of these photons. For a recorder having a quantum efficiency η, the signal-to-noise ratio becomes $C\sqrt{\eta\Phi A R t}$. Here we see the fundamental trade-off between resolution A and dose R. For a typical chest x-ray the exposure is about 50 mR, the transmission t through regions devoid of bone is about 0.05, and the quantum efficiency of the screen is about 0.25. For a linear resolution of about 1.0 mm the signal-to-noise ratio is about $400C$ or about 16 db for $C = 0.1$, a lesion exhibiting a 10% change in transmission.

The signal-to-noise ratio has been defined in terms of the incident radiation exposure and thus effectively in terms of the surface dose. The maximum dose usually occurs at the body surface. Although this is an effective measure, a number of other dose indicators are used. It is important to note, however, that each of these relates directly to the incident photon density and thus would differ only by a constant for a given body material. Integral dose, or total absorbed energy, is expressed in gram-rads, where a gram-rad is equivalent to 100 ergs. Another global measure of dose used is the surface integral exposure, which is the product of the exposure in roentgens and the total surface area exposed.

RESOLUTION CONSIDERATIONS OF THE SNR

The SNR in equations (6.9) and (6.11) is clearly proportional to the contrast or fractional change in transmission caused by the structure of interest. It should be emphasized that this contrast C is the recorded contrast. For large lesions C is unaffected by the blurring due to the finite source and the recorder resolution since this blurring merely rounds the edges of the lesion and does not change its central value. For smaller structures, however, the recorded contrast will definitely be affected. In these cases a specific definition must be assigned to C, relating to the shape of the recorded image. Hopefully, this definition will strongly relate to visualization by observers.

For example, a planar object $t(x, y)$ can be blurred due to a finite source size, providing an image as given in Chapter 4 of the form $s(x/m, y/m) ** t(x/M, y/M)$. A small lesion can have its recorded contrast considerably altered by this convolution operation. It is interesting to note that this system has an optimum source size depending on the structure being visualized. A relatively small source provides the highest recorded contrast, but with relatively few photons. As the source size increases, for a given photon density, the number of photons increases but, in general, the contrast can decrease due to the convolution operation. This, of course, is strongly dependent on the depth position of the lesion of interest.

RECORDER STATISTICS

Ideally, the number of photons per picture element transmitted through the body completely determines the resultant SNR performance. This, however, is only true with an ideal recorder. Equation (6.11) thus represents the best possible SNR for a given radiation. In practice, only a portion η of the x-ray photons are captured. Also, an additional noise source arises in the utilization of the captured photons to record the image.

The recording process most frequently used in radiography and fluoroscopy is the scintillation screen. These screens use relatively high atomic number materials, such as calcium tungstate, where the x-ray photons are stopped primarily by the photoelectric effect. In this process an electron is raised from the valence band to the conduction band. In its return to the valence band energy is radiated in the form of visible light photons. The number of light photons produced due to each captured x-ray photon is also a Poisson-distributed random variable. The total number of light photons produced is given by

$$Y = \sum_{m=1}^{M} X_m \tag{6.12}$$

where Y is a random variable representing the total number of light photons produced, M is a random variable representing the number of captured x-ray photons, and X_m is a random variable representing the gain or the number of light photons produced per x-ray photon.

Our resultant SNR will be determined by the statistics of Y, the total light photons. We will study to what extent this SNR has been reduced compared to that of equation (6.11). We do not need to know the probability density distribution of Y [Feller, 1957] since the SNR is completely determined by the mean and variance. Using probability theory, these are given by

$$E(Y) = E(M)E(X) \tag{6.13}$$

and

$$\sigma_y^2 = E(M)\sigma_x^2 + \sigma_M^2 E^2(X) \tag{6.14}$$

where E represents the expected value and σ^2 the variance. These equations are intuitively reasonable since the resultant mean would be expected to be the product of the individual means. The variance σ^2 is due to the uncertainty of the number of captured x-ray photons and the uncertainty in the number of light photons produced per x-ray photon. Since each x-ray photon produces light photons having a variance σ_x^2, the total uncertainty due to light photons is $E(M)\sigma_x^2$. Similarly, the variance in the number of captured x-ray photons σ_M^2 is subject to an average gain of $E(X)$, so that this resultant component of the variance is weighted by $E^2(X)$.

We can use these results to analyze the signal-to-noise ration of a radiographic screen to determine whether the uncertainty in the emission of light photons influences the overall performance. We let $E(X)$, the mean value of the number of light photons produced from each captured x-ray photon, be g_1, representing the average gain. It should be emphasized that g_1 includes both the generation of light photons and their attenuation or transmission loss to the point where they are used. The number of light photons captured is often limited by collection angle or transmission through a material. As with the previous analysis, when a Poisson process is attenuated by one having a probability of transmission p, the effective rate of the Poisson process is simply multiplied by p. Thus g_1 is the product of the average number of light photons produced per x-ray photon multiplied by the probability of transmission to the point of use. Since X is Poisson distributed, g_1 is also the variance of X. $E(M)$, the expected number of captured photons per picture element is, as before, ηN. Since M is also Poisson distributed, its variance is ηN. Substituting these into equations (6.13) and (6.14), we obtain

$$E(Y) = \eta N g_1 \tag{6.15}$$

and

$$\sigma_y^2 = \eta N g_1 + \eta N g_1^2. \tag{6.16}$$

The resultant signal-to-noise ratio is given by

$$\text{SNR} = \frac{CE(Y)}{\sigma_y} = \frac{C\sqrt{\eta N}}{\sqrt{1 + 1/g_1}}. \tag{6.17}$$

If an appreciable number of light photons g_1 are collected for each captured x-ray photon, the resultant statistics are essentially determined by the number of captured photons ηN.

In most processes, such as the radiographic screen–film combination, at least one additional gain stage is involved having a similar statistical model. In the second stage each light photon generates a Poisson-distributed array of events whose mean is g_2. We can characterize the sum of the recorded events in this process by W, where

$$W = \sum_{m=1}^{Y} Z_m \tag{6.18}$$

where Y, as before, is the number of captured light photons and Z_m is a random variable representing the number of events, such as developed film grains, produced for each captured light photon. Using the previous relationships, we obtain

$$\sigma_w^2 = E(Y)\sigma_z^2 + \sigma_y^2 E^2(Z)$$

$$= g_1 g_2 \eta N + (\eta N g_1 + \eta N g_1^2) g_2^2 \tag{6.19}$$

$$\sigma_w = g_1 g_2 \sqrt{\eta N} \sqrt{1 + \frac{1}{g_1} + \frac{1}{g_2}} \tag{6.20}$$

$$\text{SNR} = \frac{CE(W)}{\sigma_w} = \frac{C\sqrt{\eta N}}{\sqrt{1 + 1/g_1 + 1/g_1 g_2}}. \tag{6.21}$$

This representation can be generalized for q successive stages as

$$\text{SNR} = \frac{C\sqrt{\eta N}}{\sqrt{1 + \sum_{i=1}^{q}\left(1/\prod_{j=1}^{i} g_j\right)}}. \tag{6.22}$$

If every product term, $g_1 g_2 g_3, \ldots$, is appreciably greater than unity, the SNR will be dominated by the captured number of x-ray photons per pixel. In general, this is a desirable goal since it ensures that the minimum patient dosage is used for a given image quality.

In the particular case of radiographic film–screen cassettes, g_1, the gain of the scintillating screen, is limited by energy conservation to the ratio of the wavelength of the emitted visible light, 5000 Å, to that of the x-ray photons, 0.25 Å, which is 20,000. These screens have an energy efficiency of about 5%, about half of which escapes, resulting in a g_1 of about 0.5×10^3. This first stage does not significantly change the resultant signal-to-noise ratio. The final process, the film emulsion, is highly nonlinear and not subject to simple analysis. As a linearizing approximation, however, about 200 photons are required on the average for each developed silver grain in x-ray film; thus $g_2 \simeq 1/200$. The $g_1 g_2$ product of 2.5 does result in about a 20% degradation of the performance since $\sqrt{1 + 2/5} \simeq 1.2$.

FLUOROSCOPY

In fluoroscopy [Ter-Pogossian, 1967; Christensen et al,. 1978] the image is displayed in real time rather than being recorded on film. This allows moving structures such as the heart to be visualized. It is also widely used for the insertion of catheters to monitor their position. In the early forms of fluoroscopy the radiologist directly observed the fluorescent screen as shown in Fig. 6.3.

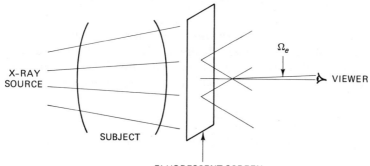

FIG. 6.3 An early fluoroscopy system.

This system has a significant noise problem due to a deteriorated light collection efficiency. Approximately 10^{-5} of the light quanta produced at the screen actually appear at the retina of the eye because of the small solid angle intercepted by the pupil Ω_e and the losses within the eye. The capture efficiency of the eye η_e is given by

$$\eta_e = \frac{T_e \Omega_e}{4\pi} = \frac{T_e A}{4\pi r^2} \tag{6.23}$$

where T_e is the light transmission to the retina, which is about 0.1, A is the pupil area and r is the distance of the eye from the screen. For the dark-adapted eye, A is about 0.5 cm², corresponding to an 8-mm pupil. The closest reasonable viewing distance is about 20 cm, resulting in maximum value of η_e of about 10^{-5}. Values of 10^{-7} to 10^{-8} are much more typical. For a typical screen the product of the screen gain of 10^3 and the retinal transfer of 10^{-5} is about 10^{-2}

at best. This reduces the signal-to-noise ratio by 10. Stated in other words, the photon flux, or radiation, required to provide an image having a signal-to-noise ratio comparable to that recorded on film would require a 100-fold increase. As fluoroscopy was normally practiced, however, the radiologist contented himself with a poorer signal-to-noise ratio. He usually employed dark-adapted vision, which increased pupil size and thus maximized the collection efficiency of light photons. In the dark-adapted state, however, the visual accuity is considerably reduced since the density of rods in the retina is considerably less than that of the cones, which dominate vision at normal light levels.

Thus early fluoroscopy was characterized by poor statistics and poor visual performance of the observer. It is the function of the image intensifier, our next topic, to both improve the statistics, in terms of available independent events per pixel, and to increase the brightness sufficiently so as to provide normal visual acuity.

IMAGE INTENSIFIER

Modern fluoroscopic systems [Ter-Pogossian, 1967; McLean and Schagen, 1979] solve these problems through the use of the image intensifier, shown in Fig. 6.4. In these devices the light from the phosphor scintillator is amplified before being utilized by the eye. A photoemissive material is placed against the scintillating phosphor. These materials have quantum efficiencies of about 10%, so that one electron is emitted for about every 10 light photons. This provides

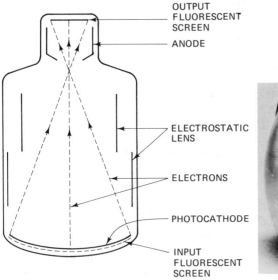

FIG. 6.4 Diagram of an x-ray image intensifier tube and a photograph of a typical intensifier. (Courtesy of the Siemens AG-Bereich Medizinische Technik.)

a g_1g_2 product of the phosphor screen and photoemitter gains of about 100, which will leave the incident signal-to-noise ratio essentially undisturbed. The emitted electrons are focused by the various electrostatic lenses and magnets so as to reproduce the incident image with good fidelity on an output phosphor screen. These electrons are accelerated to an energy of about 25 kev so as to produce about 10^3 visible photons per electron. The resultant intensified image can either be observed directly or through television or film cameras. If observed directly, the light loss to the retina of about 10^{-5} results in an overall gain $g_1g_2g_3$ of about 1, so that the signal-to-noise ratio is slightly reduced. Figure 6.5 illustrates the quantum values at different levels in the system. In a dose-

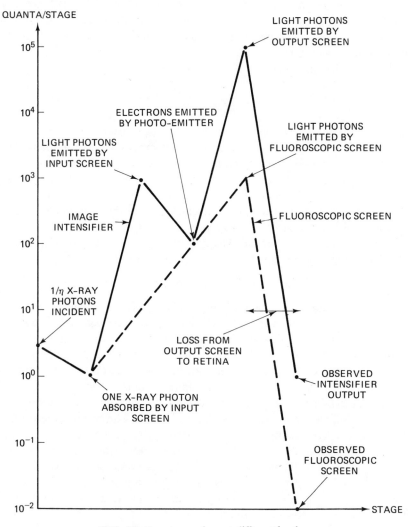

FIG. 6.5 Quantum values at different levels.

efficient system the minimum quantum levels, representing the product of the average gains, remain well above that of the absorbed x-ray photons.

The demagnification of the input image to a smaller output image in Fig. 6.4 does not affect the SNR, since it is based on events per picture element. This demagnification does, however, increase the brightness of the output image, which does affect the visual accuity of the observer. Increased brightness can ensure that the visual accuity will be dominated by the high-resolution cones rather than the rods.

With many image intensifiers the small image size precludes direct viewing so that additional devices are required. If optics are used to magnify the image, much of the image intensifier gain will be lost, thus providing either a noisy image or additional exposure. This problem can be alleviated by coupling the output of the image intensifier to a television camera tube. The optical system coupling the image intensifier output to the television camera is usually a lens. The photon collection efficiency of a lens η_0 is given by

$$\eta_0 = \frac{M^2 T}{(M + 1)^2 4f^2} \qquad (6.24)$$

where M is the magnification, f the ratio of image distance to lens diameter, and T the light transmission of the optics. This expression is essentially determined by the solid angle intercepted by the optical system. Systems that couple the image intensifier output to a television camera have f numbers of about 1.0, a magnification of about 1.0, and a transmission T of about 0.6, resulting in an η_0 of about 0.04. The quantum efficiency of the photoemitter of the television camera is about 10 %. This represents the lowest point in the quantum amplification chain, since beyond this point there are a variety of amplification mechanisms. The $g_1 g_2 g_3 g_4$ at this lowest point is given by

$$(10^3)(10^{-1})(0.04 \times 10^3)(0.1) = 4 \times 10^2$$

where g_1 is the gain of the scintillating phosphor in the image intensifier in light photons per x-ray photon, g_2 the efficiency of the photocathode in electrons per light photon, g_3 the gain of the output phosphor of the image intensifier with the loss in the optics in light photons per electron, and g_4 the quantum efficiency of the television camera photocathode in electrons per light photon. The result of 4×10^2 ensures that the signal-to-noise ratio of the system is essentially determined by the number of quanta emerging from the body and the quantum efficiency in capturing these photons. Fluoroscopy becomes an efficient process. The resolution through the many cascaded imaging structures is reduced over film radiography, although adequate for most studies. In addition to resolution, the dynamic range is reduced considerably compared to photographic film because of the television camera and the glare of the image intensifier. Figure 6.6 illustrates the quantum values at different levels in the system employing a television camera.

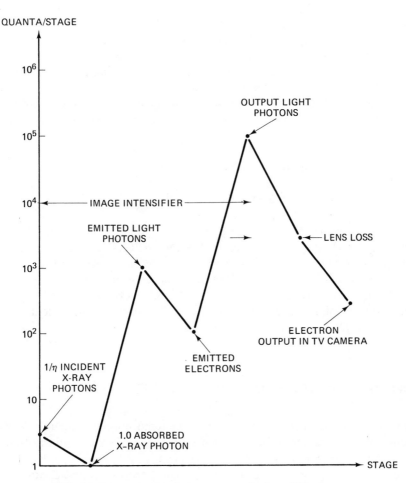

FIG. 6.6 Quantum values in a system employing a TV camera.

ADDITIVE NOISE

The additive electrical noise component in the output signal of the television camera limits the dynamic range and can represent the limiting noise component in regions of high photon transmission. Since this additive noise is independent of the Poisson noise due to the x-ray photons, the signal-to-noise ratio can be structured as

$$\text{SNR} = \frac{C\eta N}{\sqrt{N_a^2 + \eta N}} \tag{6.25}$$

where N is the average or background number of transmitted photons per element, $N = N_0 \exp(-\int \mu dz)$, C is the fractional variation of the region under study, and N_a^2 is the variance of the additive noise component. Thus N_a, the standard deviation of the additive noise, is being expressed, for convenience, as a number of photons per element.

In general, the additive noise N_a will occupy a fraction k of the dynamic range or the average value of the signal. This is conveniently expressed as

$$N_a = k\eta N. \tag{6.26}$$

The SNR in equation (6.25) can be rewritten as

$$\text{SNR} = \frac{C}{k}\sqrt{\frac{1}{1 + 1/k^2\eta N}}. \tag{6.27}$$

In cases where $k^2\eta N \gg 1$, corresponding to a high photon count and/or a relatively high additive noise fraction, the SNR reduces to C/k, the ratio of the fractional signal component to the additive noise. In this case further increases in radiation N_0 will not improve the performance, since it is being dominated by the additive noise of the system. In the other extreme, where $k^2\eta N \ll 1$, the additive noise is negligible and the SNR returns to the dose-dependent case of $C\sqrt{\eta N}$.

SNR OF THE LINE INTEGRAL

We have thus far dealt solely with the signal and noise considerations of the transmitted photon intensity. This transmitted intensity is a nonlinear function of the line integral of the attenuation coefficient $N = N_0 \exp(-\int \mu dz)$. In more recent x-ray imaging systems employing digital processing and electronic displays, the line integral g itself is calculated and displayed through a log operation as given by

$$g = \ln N_0 - \ln N = \int \mu dz \tag{6.28}$$

where the detector efficiency has been assumed unity.

In this case our SNR is defined as

$$\text{SNR} = \frac{\epsilon \bar{g}}{\sigma_g} \tag{6.29}$$

where ϵ is the fractional change in the average value of $\int \mu dz$ caused by the region of interest and σ_g is the standard deviation.

To evaluate the mean and standard deviation of the line integral g, we make use of the fact that g is a function of the random variable X [Papoulis, 1965], which is known to be Poisson distributed. We expand the function g

$$C \frac{\Delta I}{\sigma_c} \Rightarrow \epsilon \, (\text{contrast}) \sqrt{N}$$

in a Taylor series about N as given by

$$g(X) = \ln N_0 - \ln X$$
$$= g(N) + g'(N)(X - N) + g''(N)\frac{(X - N)^2}{2} + \ldots + g^{(n)}(N)\frac{(X - N)^n}{n!}$$

(6.30)

where X is a random variable representing the number of transmitted photons per element and N is its mean value.

Using the general relationship

$$E[g(X)] = \int g(X)p(X)dx \tag{6.31}$$

we can use the series to evaluate the various statistical averages of g. Inserting (6.30) into (6.31) and integrating, we obtain

$$E[g(X)] = g(N) + g''(N)\frac{\sigma^2}{2} + \ldots \tag{6.32}$$

where σ^2, as before, is the variance of the number of photons per element given by N. Evaluating the second derivative of the log, we obtain

$$E[g(X)] = \ln\left(\frac{N_0}{N}\right) + \frac{1}{2N} + \ldots . \tag{6.33}$$

Since the number of counts per element exceeds 10^6 in most circumstances, the expected value of the line integral can simply be given by $\ln(N_0/N)$, the same as our deterministic approximation of equation (6.28).

For the variance of the line integral, σ_g^2, we use equation (6.31) to evaluate $E[g^2(X)]$ as given by

$$E[g^2(X)] = g^2(N) + [g'(N)^2 + g(N)g''(N)]\sigma^2 + \ldots . \tag{6.34}$$

Using the relationship from equation (2.49) yields

$$\sigma_g^2 = E[g^2(X)] - E^2[g(X)]$$
$$\simeq [g'(N)]^2\sigma^2 - \left[g''(N)\frac{\sigma^2}{2}\right]^2 + \ldots$$
$$\simeq \frac{1}{N} + \text{terms of degree } \frac{1}{N^2} \text{ and higher}$$
$$\simeq \frac{1}{N}. \tag{6.35}$$

Thus the signal-to-noise ratio is given, from equation (6.29), by

$$\text{SNR} = \epsilon \ln\left(\frac{N_0}{N}\right)\sqrt{N} \tag{6.36}$$

or, including the detection efficiency,

$$\text{SNR} = \epsilon \ln\left(\frac{\eta N_0}{\eta N}\right)\sqrt{\eta N}$$
$$= \epsilon\left(\int \mu dz\right)\sqrt{\eta N}. \tag{6.37}$$

Two appropriate assumptions have been made in this derivation, which should be pointed out for completeness. First, to use the Taylor series expansion, we must assume that the probability of collecting zero counts is zero; otherwise, the log function blows up. This is clearly a reasonable assumption, given an average of $> 10^6$. Second, the statistics of the number of incident photons per element, N_0, has been ignored. This is again reasonable since N_0 is generally a multiplicative factor of $> 10^2$ greater than N, so that its relative statistical variation can be neglected. Also, monitoring detectors are often used to measure N_0, thus minimizing the effect of its fluctuations.

The expression for ϵ, the fractional change in the line integral, can be structured in terms of the contrast C, as given by

$$C = \frac{\Delta N}{N} = \frac{N_0 e^{-\int \mu dz} - N_0 e^{-(1+\epsilon)\int \mu dz}}{N_0 e^{-\int \mu dz}} \simeq \epsilon \int \mu dz. \qquad (6.38)$$

Therefore, logarithmic processing of the image, to provide a display of the line integral of the attenuation coefficient, results in the same SNR as the direct intensity presentation as given by

$$\text{SNR} = \frac{\epsilon \bar{g}}{\sigma_g} = \frac{C\bar{I}}{\sigma_I} = C\sqrt{\eta N}. \qquad (6.39)$$

SCATTER

The most significant additive noise component is that of scatter [Ter-Pogossian, 1967]. The attenuation mechanism that dominates radiography is Compton scattering. Depending on the energy used and the atomic number of the material, the attenuation mechanism is divided between absorption and scatter. In the soft-tissue regions, which occupy most of the body, scattering is the dominant mechanism. Unfortunately, many of the scattered photons reach the screen, representing additive noise.

Scatter, as an additive noise component, has two deleterious effects, a loss of contrast and an increase in noise. The first is a deterministic phenomenon where the scatter produces an added intensity I_s which adds to the transmitted intensity I_t. The resultant contrast C_r due to a transmission contrast C is given by

$$C_r = \frac{CI_t}{I_t + I_s} = \frac{C}{1 + I_s/I_t} \qquad (6.40)$$

where $(1 + I_s/I_t)^{-1}$ is the contrast reduction factor. This reduction is clearly evident in regions of low transmission where I_t becomes comparable to I_s. In addition to its deterministic effect, the scattered photons produce an additional noise component in the counting statistics. If N represents the transmitted photons per picture element and N_s the scattered photons per picture element, their variances add in the detector because of independence. The resultant signal-to-noise ratio becomes

$$\text{SNR} = \frac{C\eta N}{\sqrt{\eta N + \eta N_s}} = \frac{C\sqrt{\eta N}}{\sqrt{1 + N_s/N}}. \tag{6.41}$$

Scatter causes a serious deterioration of performance if nothing is done to reduce it. This scatter reduction process must somehow distinguish between the scattered and transmitted photons. One mechanism, as described in Chapter 3, is to make use of the photon energy loss that takes place in the process of Compton scattering. Unfortunately this can only be used for the case of mono-energetic sources, since, with broadband sources, the transmitted and scattered photons have overlapping energy spectra. This procedure of scatter elimination through spectral analysis is thus only used in nuclear medicine, where the isotope sources are monoenergetic and of sufficiently high initial energy to cause a significant energy loss. The other mechanism that distinguishes scatter is that of direction. Transmitted photons all appear to come from the source. Scattered photons can thus be minimized by collimating structures that are aimed at the source. These structures, known as *grids*, absorb many of the scattered photons because they arrive at angles other than that determined by the position of the source. They will be analyzed in some detail following the development of a model to analyze the amount of scatter.

SCATTER ANALYSIS

As illustrated in Fig. 6.7, the area under study is a cylinder of length L and radius R which is assumed to be homogeneous with an attenuation coefficient of μ. The incident x-rays are assumed to be parallel having a photon intensity of n_0 photons/cm^2. Thus the intensity reaching the incremental section at plane z has a photon density given by $n(z)$. The incremental density of scattered photons generated in this section is given by

$$dn_s(z) = n(z)\mu_s dz = n_0 e^{-\mu z}\mu_s dz \tag{6.42}$$

where μ_s is the Compton scattering portion of the attenuation coefficient.

We now consider what fraction of the scattered photons will reach the detector. If the scatter were perfectly isotropic, half of the scattered photons would be in the direction of the detector and the other half toward the source. In the diagnostic energy range there is a slight departure from isotropy, where the fraction k that is scattered in the forward direction [Klein and Nishima, 1929] is approximated by

$$k = 0.52 + \frac{0.07E(\text{kev})}{80}. \tag{6.43}$$

Thus at a typical average photon energy of 40 kev, $k = 0.55$.

If we assume that the collection angle at each point in the detector plane is constant at $\Omega(z)$, the fraction of scattered photons $F(z)$ that reach the detector is given by

$$F(z) = ke^{-\mu(L-z)}\frac{\Omega(z)}{2\pi} \tag{6.44}$$

where k is the fraction scattered forward, $e^{-\mu(L-z)}$ is the fraction transmitted through the material, and $\Omega(z)/2\pi$ is the fraction subtended by the solid angle of the detector. If these scattered photons were the only ones produced, neglecting secondary scattering processes, the total scattered photon density at the detector, n_s', would be given by

$$n_s' = \int n(z)\mu_s F(z)dz. \tag{6.45}$$

However, this analysis has neglected a variety of secondary processes. These include the scattered photons within the solid angle which have experienced an additional scattering event but remain within the solid angle. The analysis has also neglected photons which were scattered outside the solid angle, which, as a result of a subsequent scattering event, are scattered back into the solid angle. We can summarize this process by a multiplicative buildup factor B providing a total scattering density at the detector n_s given by

$$n_s = \int_0^L Bn(z)\mu_s F(z)dz$$
$$= Bn_0 ke^{-\mu L}\mu_s \int_0^L \frac{\Omega(z)}{2\pi} dz. \tag{6.46}$$

We first evaluate $\Omega(z)$ and its integral for the cylindrical geometry of Fig. 6.7 and then evaluate the buildup factor B. Using r, θ polar coordinates at the

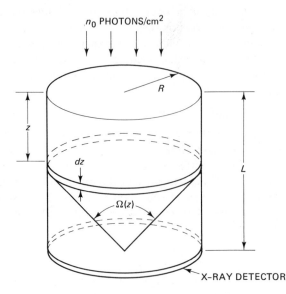

n_0 PHOTONS/cm^2

R

z

dz

$\Omega(z)$

L

X-RAY DETECTOR

FIG. 6.7 Model for scatter analysis.

detector plane in Fig. 6.7, we have

$$d\Omega = \frac{dA \cos \alpha}{r^2 + (L - z)^2} = \frac{r\,dr\,d\theta(L - z)}{[r^2 + (L - z)^2]^{3/2}} \qquad (6.47)$$

where $dA \cos \alpha$ is the incremental area element in the plane of the section with α being the angle to the axis of the cylinder. Integrating, we obtain

$$\Omega(z) = (L - z) \int_0^R \int_0^{2\pi} \frac{r\,dr\,d\theta}{[r^2 + (L - z)^2]^{3/2}}$$

$$= 2\pi \left[1 - \frac{L - z}{\sqrt{R^2 + (L - z)^2}} \right]. \qquad (6.48)$$

The integral in equation (6.46) is then given by

$$\int_0^L \frac{\Omega(z)}{2\pi}dz = L + R - \sqrt{L^2 + R^2} \qquad (6.49)$$

with the scattered photon density at the detector given by

$$n_s = Bn_0 k e^{-\mu L} \mu_s (L + R - \sqrt{L^2 + R^2}). \qquad (6.50)$$

The multiplicative buildup factor B accounts for those scattered photons not collected after the first scattering. If they scatter again, the fraction collected can be structured exactly as the previous analysis—hence the multiplicative nature of the factor. A simplified estimate of B would therefore be the estimated number of scattering events as a photon travels to the screen. We first calculate the mean distance in the z direction to a scattering interaction of an entering photon. The estimate of B is then the total depth L divided by this mean distance.

Assume that a photon is moving toward the detector at an angle α to the cylindrical axis. The distribution function of a scattering interaction along its path or probability that the photon will reach a distance r before interacting is given by

$$F(r) = 1 - e^{-\mu_s r}. \qquad (6.51)$$

This corresponds to the probability that the photon will reach a distance $z/\cos \alpha$ as given by the conditional distribution function

$$F(z\,|\,\alpha) = 1 - \exp\left(\frac{-\mu_s z}{\cos \alpha}\right). \qquad (6.52)$$

Differentiating provides the probability density function,

$$p(z\,|\,\alpha) = \frac{\mu_s}{\cos \alpha} \exp\left(\frac{-\mu_s z}{\cos \alpha}\right). \qquad (6.53)$$

The conditional mean in the z direction, given the angle α, becomes

$$E(z\,|\,\alpha) = \int zp(z\,|\,\alpha)dz$$

$$= \frac{\mu_s}{\cos \alpha} \int_0^\infty z \exp\left(\frac{-\mu_s z}{\cos \alpha}\right)dz = \frac{\cos \alpha}{\mu_s}. \qquad (6.54)$$

The mean or average distance to an interaction in the z direction is obtained by integrating the conditional expectation multiplied by the probability density of α. Since the scatter is isotropic, it has equal probability of appearing in any two-dimensional angular interval. This corresponds to $p(\alpha) = \sin \alpha$, where α varies from 0 to $\pi/2$. The resultant mean distance is

$$E(z) = \int E(z \mid \alpha) p(\alpha) d\alpha$$

$$= \int_0^{\pi/2} \frac{\cos \alpha}{\mu_s} \sin \alpha \, d\alpha$$

$$= \int_0^{\pi/2} \frac{1}{2\mu_s} \sin 2\alpha \, d\alpha = \frac{1}{2\mu_s}. \tag{6.55}$$

The average number of interactions along the entire length L is given by

$$B = \frac{L}{E(z)} = 2\mu_s L \tag{6.56}$$

with the scattered photon density at the detector given by

$$n_s = n_0 e^{-\mu L} \mu_s^2 L 2k(L + R - \sqrt{L^2 + R^2}). \tag{6.57}$$

We can use this result to find the signal-to-noise ratio from (6.44) as given by

$$\text{SNR} = \frac{C\sqrt{\eta n_0 e^{-\mu L} A}}{\sqrt{1 + n_s/n_0 e^{-\mu L}}} \tag{6.58}$$

where A is again the area of a picture element. It is convenient to structure the SNR as

$$\text{SNR} = \frac{C\sqrt{\eta n_0 e^{-\mu L} A}}{\sqrt{1 + \psi}} \tag{6.59}$$

where ψ is the ratio of scattered-to-transmitted photon density at the detector as given by

$$\psi = \frac{n_s}{n_0 e^{-\mu L}} = \mu_s^2 L 2k(L + R - \sqrt{L^2 + R^2}). \tag{6.60}$$

Using typical values of a body section where $L = 20$ cm, $R = 10$ cm, with $\mu_s = 0.2$ cm^{-1}, and $k = 0.5$, we have a ψ of approximately 6.1.

This degree of scatter is clearly unacceptable. It would require a corresponding increase of greater than 6 to 1 in radiation dose to restore the original SNR. In addition, the contrast, especially in areas of low transmission, is seriously deteriorated. It is therefore obvious that scatter reduction is required.

SCATTER REDUCTION THROUGH SEPARATION

Because of the many multiple scatter processes, the scattered photons emitted from the volume in Fig. 6.7 can be assumed to be approximately isotropic, covering a solid angle of 2π steradians. If the detector is separated from the object by some distance s, the number of received scattered photons will be reduced while those transmitted will be unaffected. The solid collecting angle $\Omega(s)$ between the detector and the emitted photons from the object has the same form as derived for $\Omega(z)$ in equation (6.48) and is given by

$$\Omega(s) = 2\pi \left(1 - \frac{s}{\sqrt{R^2 + s^2}}\right). \tag{6.61}$$

Therefore, the fraction of scattered photons collected by the screen is given by $\Omega(s)/2\pi$. A plot of this collection factor versus s/R, the ratio of the separation to the object radius, is shown in Fig. 6.8. As can be seen, for separations com-

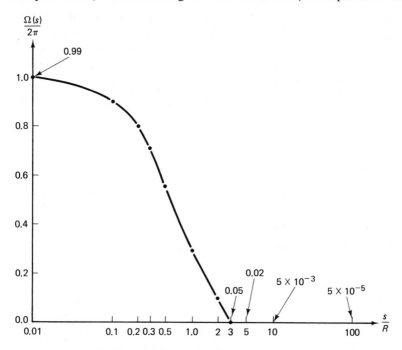

FIG. 6.8 Scatter collection fraction vs. separation/radius.

parable to the object size, the scatter collection fraction is about 0.3, leaving the resultant scatter intensity larger than the transmitted intensity for the object previously considered. For relatively large separations, however, this reduction factor becomes appreciable. However, at these larger separations, we experience reduced resolution due to the finite source size as studied in Chapter 4. With separation the SNR is given by

$$SNR = \frac{C\sqrt{\eta n_0 e^{-\mu L} A}}{\sqrt{1 + \psi\Omega(s)/2\pi}}. \tag{6.62}$$

Equation (6.62) was derived with the assumption of a parallel beam. Although this assumption gives reasonably accurate results for the previous analysis, it does give an optimistic result in the case of separation. In the case of a divergent beam, the detected photon density due to the transmitted signal photons will decrease due to the geometric considerations. The scatter, however, is unaffected by the beam divergence since it essentially represents an isotropic source. Thus the SNR is reduced from that shown in equation (6.62).

SCATTER REDUCTION USING GRID STRUCTURES

The scatter can be reduced, with a small loss of transmitted photons, by using a collimated structure which is parallel to the direction of the transmitted beam [Ter-Pogossian, 1967]. As shown in Fig. 6.9, the grid consists of an array of thin absorptive strips which are parallel for the case of a collimated x-ray beam and are pointed toward the source for a diverging beam. The latter is called a *focused grid* since it is designed for a specific source to detector distance.

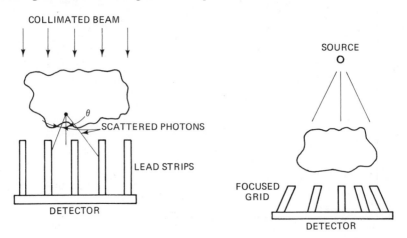

FIG. 6.9 Scatter-reducing grids.

Since the analysis is unduly complex using a focused grid, we will study the collimated grid and make the reasonably accurate assumption that the scatter-reduction performance will be essentially identical in both cases. In evaluating the performance of a grid structure, we calculate its transmission as a function of scatter angle in one dimension, $T(\theta)$. The reduction factor R_s of the scattered photons is then given by

$$R_s = \frac{\int_{-\pi/2}^{\pi/2} n_s(\theta)T(\theta)d\theta}{\int_{-\pi/2}^{\pi/2} n_s(\theta)d\theta} = \frac{2}{\pi}\int_0^{\pi/2} T(\theta)d\theta \qquad (6.63)$$

where $n_s(\theta)$ is assumed uniformly distributed over π radians and $T(\theta)$ is even. The function of the grid structure is thus to provide as low as possible an integrated $T(\theta)$ without unduly attenuating the desired transmitted photons. If η_t is the grid efficiency for transmitted photons, the improvement in the ratio of transmitted to scattered photons at the detector is given by η_t/R_s. $T(\theta)$ and R_s are evaluated using Fig. 6.10.

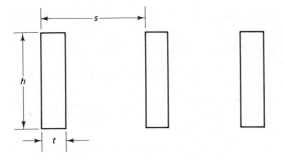

FIG. 6.10 Dimensions of grid strips.

Scatter reduction is achieved by providing a relatively high transmission to the collimated desired photons, and a relatively low transmission to the isotropic scattered photons.

We first subdivide $T(\theta)$ into specific angular regions which have uniform properties. Within each angular region, at each θ, we use the fact that each ray has a uniform probability distribution of occupying each lateral position. We thus further subdivide each angular region into regions having different attenuation mechanisms and apportion each region uniformly. In those regions having attenuations which vary with ray translation, we integrate to find the mean attenuation. For example, for rays having angles to the normal $0 < \theta < \tan^{-1}(t/h)$, the rays are either totally in the metal strip, partially in the metal strip, or unattenuated. $T(\theta)$ is given by

$$\underset{0<\theta<\tan^{-1}(t/h)}{T(\theta)} = \frac{t - h\tan\theta}{s} e^{-\mu h/\cos\theta} \quad \text{(totally in metal)}$$

$$+ \frac{2h\tan\theta}{s} \frac{1}{h\tan\theta} \int_0^{h\tan\theta} e^{-\mu x/\sin\theta}\, dx \quad \text{(partially in metal)}$$

$$+ \frac{s - t - h\tan\theta}{s} \times 1 \quad \text{(unattenuated)} \tag{6.64}$$

where μ is the attenuation coefficient of the metal strip. The first term has the factor $(t - h\tan\theta)/s$, indicating the fraction of the period s that the rays are totally in the length of the metal strip and experience the attenuation $\exp(-\mu h/\cos\theta)$. In the second term the duty cycle is $(2h\tan\theta)/s$, and the attenuation term is the integrated average over the different parts of the metal strip traversed by the ray. The third term is the fraction of the duty cycle, at this angular range, over which the rays do not strike the metal strip, and thus are unattenuated. Collecting terms and performing the integration, we obtain

$$\underset{0<\theta<\tan^{-1}(t/h)}{T(\theta)} = \frac{1}{s}\left[(t - h\tan\theta)e^{-\mu h/\cos\theta} + \frac{2\sin\theta}{\mu}(1 - e^{-\mu h/\cos\theta}) \right. $$

$$\left. + s - t - h\tan\theta \right]. \qquad (6.65)$$

Similarly, we find the transmission in the next angular region $\tan^{-1}(t/h) < \theta < \tan^{-1}[(s - t)/h]$, where the rays are either unattenuated, pass through the entire strip width, or pass through a part of the strip as given by

$$\underset{\tan^{-1}(t/h)<\theta<\tan^{-1}[(s-t)/h]}{T(\theta)} = \frac{s - t - h\tan\theta}{s} \times 1 \quad \text{(unattenuated)}$$

$$+ \frac{h\tan\theta - t}{s}e^{-\mu t/\sin\theta} \quad \begin{pmatrix}\text{passing through entire}\\ \text{strip width}\end{pmatrix}$$

$$+ \frac{2t}{s}\frac{1}{t}\int_0^t e^{-\mu x/\sin\theta}\,dx \quad \text{(partially in strip)}$$

$$= \frac{1}{s}\left[s - t - h\tan\theta + (h\tan\theta - t)e^{-\mu t/\sin\theta} \right.$$

$$\left. + \frac{2\sin\theta}{\mu}(1 - e^{-\mu t/\sin\theta}) \right]. \qquad (6.66)$$

In the next angular region rays either pass through the entire strip width or pass partially through a strip as given by

$$\underset{\tan^{-1}[(s-t)/h]<\theta<\tan^{-1}(s/h)}{T(\theta)} = \frac{h\tan\theta - t}{s}e^{-\mu t/\sin\theta} \quad \text{(through entire strip width)}$$

$$+ \frac{s + t - h\tan\theta}{s}\frac{1}{s - h\tan\theta}\int_{h\tan\theta-(s-t)}^t e^{-\mu x/\sin\theta}\,dx$$

$$\text{(partially in strip)}$$

$$= \frac{1}{s}\left[(h\tan\theta - t)e^{-\mu t/\sin\theta} + \frac{(s + t - h\tan\theta)\sin\theta}{s - h\tan\theta} \right.$$

$$\left. \times (e^{-\mu/\sin\theta[h\tan\theta-(s-t)]} - e^{-\mu t/\sin\theta}) \right]. \qquad (6.67)$$

This procedure is continued until $T(\theta)$ becomes vanishingly small as the large-angle rays go through many metal strips and experience large attenuation. The resulting $T(\theta)$ is then integrated to find the effectiveness of the grid.

The calculations can be significantly simplified by igonring the finite width of the metal strips but considering their exact attenuation. Thus a ray passing through a lead strip of thickness t at θ would have a transmission $e^{-\mu t/\sin\theta}$. However, because of the zero thickness approximation, we can ignore those rays that pass partially through the strip. We thus use a model of zero-thickness strips of height h and separation s whose attenuation behaves as if its thickness were t. In that case we get either zero attenuation or the complete attenuation of one or more strips as given by

$$
T(\theta) = \begin{cases}
\dfrac{h \tan \theta}{s} e^{-\mu t/\sin \theta} + \dfrac{s - h \tan \theta}{s}, & 0 < \theta < \tan^{-1}\left(\dfrac{s}{h}\right) \\[2ex]
\dfrac{2s - h \tan \theta}{s} e^{-\mu t/\sin \theta} + \dfrac{h \tan \theta - s}{s} e^{-2\mu t/\sin \theta}, & \\[2ex]
\qquad \tan^{-1}\left(\dfrac{s}{h}\right) < \theta < \tan^{-1}\left(\dfrac{2s}{h}\right) & (6.68) \\[2ex]
\dfrac{3s - h \tan \theta}{s} e^{-2\mu t/\sin \theta} + \dfrac{h \tan \theta - 2s}{s} e^{-3\mu t/\sin \theta}, & \\[2ex]
\qquad \tan^{-1}\left(\dfrac{2s}{h}\right) < \theta < \tan^{-1}\left(\dfrac{3s}{h}\right).
\end{cases}
$$

Or, in general,

$$
T(\theta) = \frac{1}{s}\{[(n + 1)s - h \tan \theta]e^{-n\mu t/\sin \theta} + (h \tan \theta - ns)e^{-(n+1)\mu t/\sin \theta}\},
$$

$$
\tan^{-1}\left(\frac{ns}{h}\right) < \theta < \tan^{-1}\left[\frac{(n + 1)s}{h}\right] \qquad (6.69)
$$

where n is an integer that takes on values from zero to infinity.

The transmission efficiency η_t for the dimensions shown in Fig. 6.10 is given by $T(0) = (s - t)/s$, assuming that the rays are completely stopped by passing through the length of a strip. Often, however, filler material is used in the space between the metal strips for structural purposes. Under these conditions η_t is given by

$$
\eta_t = \frac{s - t}{s} e^{-\mu_f h} \qquad (6.70)
$$

where μ_f is the attenuation coefficient of the filler material. Normally, relatively low atomic number materials such as plastics and aluminum are used as filler to minimize the attenuation. Except for the use of filler materials, in theory, the lead strips could be made arbitrarily high for increased scatter reduction at no loss in transmission efficiency. When filler materials are used, $T(\theta)$ is modified by a multiplicative factor $e^{-\mu_f h/\cos \theta}$, ignoring the small path through the metal grids.

The SNR using a scatter reducing grid is given by

$$
\text{SNR} = \frac{C\sqrt{\eta \eta_t n_0 e^{-\mu L} A}}{\sqrt{1 + R_s \psi / \eta_t}} \qquad (6.71)
$$

where, as before, η is the quantum efficiency of the detector, n_0 is the incoming photon density, $e^{-\mu L}$ is the transmission of the body, A is the area of a picture element, ψ is the ratio of scattered to transmitted photons at the detector, and R_s is the fraction of the scattered photons passed by the grid.

Using equation (6.69), we can calculate the performance of some typical grid structures. The curves of $T(\theta)$ are shown in Fig. 6.11. In general, the values of $(R_s/\eta_t)\psi$, the ratio of scattered to transmitted photons following the grid, is

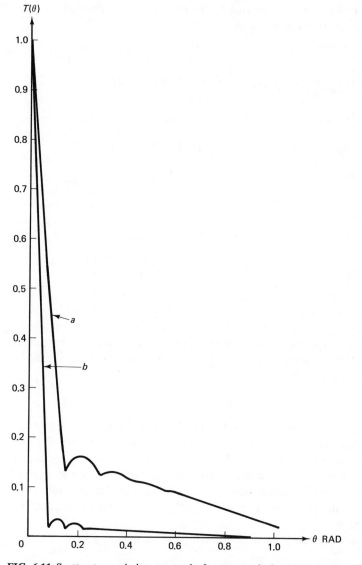

FIG. 6.11 Scatter transmission vs. angle for two typical grid structures using lead, where $\mu = 60\ \text{cm}^{-1}$. Grid (a): $s = 0.3$ mm, $h = 2.0$ mm, and $t = 0.05$ mm, resulting in $R_s = 0.106$. Grid (b) has the same parameters except that $h = 4.0$ mm, resulting in $R_s = 0.034$.

of the order of unity. This can, of course, be considerably more in a region of high attenuation, such as in the path of a large bone, where the transmitted photons are reduced and the scattered photons from the volume are relatively unaffected.

Although the grids provide reasonable performance, there are clearly many clinical situations where they are inadequate and the remaining scatter continues to limit visualization. Significant research is under way on a variety of improved scatter-reducing mechanisms.

One approach is the sequential irradiation of the body with translated slits [Sorenson and Nelson, 1976] as illustrated in Fig. 6.12. Here two translated

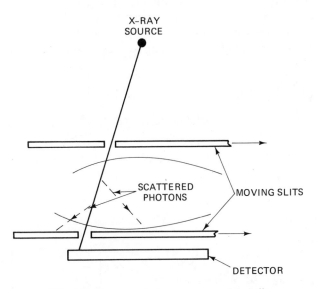

FIG. 6.12 Scatter reducing, using translated slits.

slits allow only a sheet beam to pass through the body at any one time. The slit collimators on either side of the body must be translated at different speeds to ensure that the transmitted beam is not intercepted. Most of the scattered photons will fail to reach the detector, as shown. Although these systems have performed well, they do have the practical difficulty of requiring increased average power output from the x-ray tube. Multiple-slit systems mediate this problem. Other configurations using rotating slits have also been studied because of the more convenient geometry.

LINEAR DETECTOR ARRAYS

Another version of Fig. 6.12 is the use of a one-line detector array in place of the second slit. This linear array follows the translating sheet beam, achieving the desired scatter reduction since the only sensitive area is the line itself. This one-line array has a number of added desirable features since it can be made using high-quantum-efficiency electronic detectors using individual scintillators and photodetectors. Thus not only does η approach unity, but the resultant signal is sufficiently high to override any subsequent electronic noise. The effects of neglible scatter, high efficiency, and the lack of subsequent noise processes provide a signal-to-noise ratio determined almost solely by the trans-mitted photons as given by

$$\text{SNR} \simeq C\sqrt{\Phi ART} \tag{6.72}$$

which essentially approaches the ultimate that is achievable. These linear arrays are used in conjunction with the computerized tomography scanners where they provide both cross-sectional and projection images. In this context the output signals are digitized to facilitate processing by digital computers and subsequent storage and display.

PROBLEMS

6.1 (a) Neglecting scatter and assuming that $\eta = 1$, find the SNR for imaging the thicker region where $\mu t \ll 1$ (Fig. P6.1).

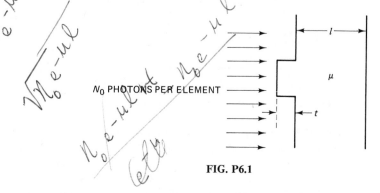

N_0 PHOTONS PER ELEMENT

FIG. P6.1

(b) Assuming that the energy dependence of μ is given by $\mu = Ae^{-B\varepsilon}$, find the energy that maximizes the SNR.

6.2 An x-ray transparency $t = a + b \cos 2\pi f_0 x$ at a depth z is imaged using an $L \times L$ rectangular source having an intensity of n photons per unit area a distance d from the recorder. Find the signal-to-noise ratio of the resul-tant image, which is defined as the ratio of the peak amplitude of the sinusoid to the standard deviation of the average value. The area of each pixel A has a negligible effect on the ability to resolve the sinusoidal image.

6.3 A circular disk x-ray source of radius r_1 emits n photons/unit area during the exposure time (Fig. P6.3). A cylindrical lesion of thickness t, radius r_2, and attenuation coefficient μ_2 is within a slab of thickness L where $L \gg t$.

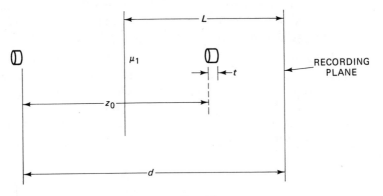

FIG. P6.3

(a) Neglecting obliquity factors calculate the signal-to-noise ratio using a recorder having an efficiency η and a resolution element of area A. The size of the signal is defined as the background minus the value at the center of the lesion image.

(b) At what ratio of r_1 to r_2 is the signal-to-noise ratio a maximum?

6.4 The cross section of an object is shown in Fig. P6.4, where the desired information is represented by the small structure of width w.

(a) Calculate the SNR of the x-ray photons emerging from the object where the size of a resolution element is A cm^2.

(b) Calculate the SNR of the recording where the phosphor screen has a capture efficiency of η and produces L light photons per x-ray photon, and

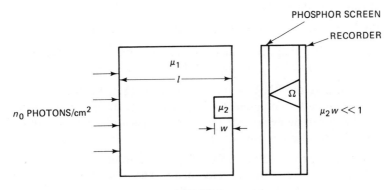

FIG. P6.4

R events per light photon are recorded. Due to critical-angle considerations, the recorder receives emitted light photons only over a solid angle Ω.

6.5 Equation (6.60) provides ψ, the ratio of scattered to transmitted photons for the cylindrical geometry of Fig. 6.7. Calculate ψ for an object having a square cross section $W \times W$ with the same length L. You can leave your answer in integral form.

6.6 A circular source of radius r_s emits n_0 photons per unit area (Fig. P6.6). A planar object of radius R has a transmission t with an opaque center of radius r_0. The image signal is defined as the difference in photons per pixel, with a pixel area A, between the background and the center of the image of the opaque disk at the center. Ignore all obliquity considerations and assume that $\eta = 1$.

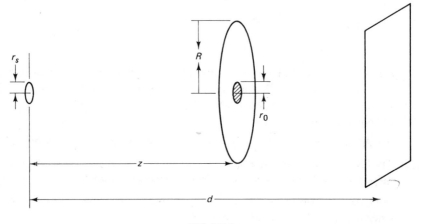

FIG. P6.6

(a) Find the value of r_s that provides the maximum SNR with the object at some specific plane z_0.

(b) Using this value of r_s, find the SNR versus depth z.

(c) Assuming that the planar object produces p isotropically scattered photons per incident photon, find the SNR versus depth z.

6.7 (a) Calculate the SNR of the lesion image (Fig. P6.7) assuming that the Compton scatter component of μ_0 is μ_s. The difference in scatter due to the lesion can be neglected. The detector has a capture efficiency of η and a resolvable element area of A.

(b) Calculate the modified SNR using a grid structure having a transmission $T(\theta)$ placed against the detector plane. The separation s is eliminated in this case with the detector against the object.

$$\psi = \frac{n_s}{}$$

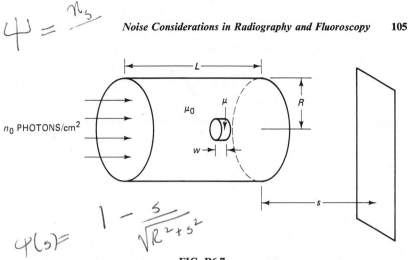

$$\psi(s) = 1 - \frac{s}{\sqrt{R^2 + s^2}}$$

FIG. P6.7

6.8 A sinusoidally modulated x-ray image is recorded by a one-sided screen–film system as shown in Fig. P6.8. Find the recorded SNR where the signal is defined as the peak of the sinusoid and the noise as the standard deviation of the average background. On the average the screen produces l light photons per x-ray photon, t of which are transmitted to the emulsion, where r of the events are recorded. The pixel area is A, which is assumed to have a negligible effect on the resolution.

$$1 + \frac{n_s}{n_t}$$

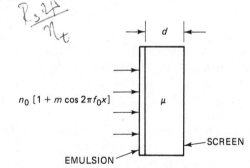

$n_0 [1 + m \cos 2\pi f_0 x]$ μ

SCREEN

EMULSION

FIG. P6.8

6.9 In the analysis of the scatter-reducing grid, the radiation from a point x-ray source was approximated as a parallel beam entering a parallel grid with a transmission efficiency $\eta_t = T(0) \simeq (s - t)/s$. Calculate η_t for a uniformly emitting $W \times W$ source a distance d from the detector. The grid has a height h, period s, thickness t, and attenuation coefficient μ. Assume $W/d < t/h$. The angles involved are sufficiently small such that $\theta \simeq \sin \theta \simeq \tan \theta$ and $\cos \theta \simeq 1$.

7

Tomography

In this chapter we consider systems that provide important "tomographic" or three-dimensional capability. The tomogram is effectively an image of a slice taken through a three-dimensional volume. Ideally, it is free of the effects of intervening structures, thus providing a distinct improvement in the ability to visualize structures of interest.

In single-projection radiography the resultant image is the superposition of all the planes normal to the direction of propagation. In essence the system has infinite depth of focus, although, as was shown in Chapter 4, the finite source size causes planes closer to the recorder to have better resolution. Ignoring this effect and assuming parallel rays, the recorded image is given by

$$I_d(x, y) = I_0 \exp\left[-\int \mu(x, y, z)dz\right]. \tag{7.1}$$

This integration over z often prevents a suitable diagnosis of the characteristics of a section at a given depth plane. Since all other planes are superimposed, the subtle contrast variations of the desired plane are often obscured. This is particularly true in studies of lung lesions where the superimposed rib structures obscure the visualization.

MOTION TOMOGRAPHY

Until very recently the only method of isolating a view of a single plane was motion tomography [Meredith and Massey, 1977], as shown in Fig. 7.1. The source and the recorder are moved in opposite directions. As shown, one plane in the object remains in focus while all others have their images blurred. The nature and degree of the deblurring is determined by the distance of each plane from the focused plane and by the extent and type of motion of the source and film. These systems are often classified by the type of motion undertaken, such as linear, circular, and hypercycloidal tomography. The mechanisms that accomplish these motions are quite elaborate since they must be both accurate and rapid, so that the motions can be completed in a few-second breath-holding interval.

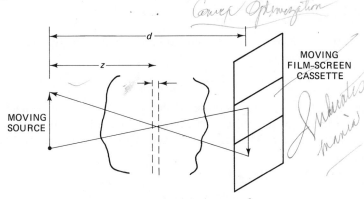

FIG. 7.1 Motion tomography.

In general, the source undergoes a specific motion in a plane parallel to the recorder plane. The path as a function of time can be characterized as $g(x, y, t)$. The path, in general, is a two-dimensional delta line function which defines the motion of the source. From this motion and the corresponding motion of the film, we can calculate the resultant impulse response.

Using the source motion of $g(x, y, t)$, we immediately see that the resultant path on a stationary film using a pinhole transparency is $g(x/m, y/m, t)$ by straightforward geometry. Thus the path due to source motion experiences the same magnification, $m = -(d - z)/z$, as did the source image in Chapter 4. As indicated, to provide a tomographic plane, the film is moved in a scaled version of the source motion. Whenever the source is displaced from the axis at a point (x, y), the film center is positioned at $-kx, -ky$, where k is a positive number representing the scaling of the film motion. Thus the center of the film traverses a path $g(x/-k, y/-k, t)$.

As the film moves, the resultant path of a beam on the film moves in the opposite direction with respect to film coordinates. Thus the resultant path incident on the film due to the two motions is the impulse response, which is given by

$$h(x_d, y_d, t) = Bg\left(\frac{x_d}{k+m}, \frac{y_d}{k+m}, t\right) \qquad (7.2)$$

where $(k+m)$ is the total magnification due to both motions and B is a normalizing constant that will be subsequently evaluated. The integrated impulse response is therefore

$$h(x_d, y_d) = B \int g\left(\frac{x_d}{(k+m)}, \frac{x_y}{(k+m)}, t\right) d(vt)$$

$$= Bf\left(\frac{x_d}{k+m}, \frac{y_d}{k+m}\right) \qquad (7.3)$$

where $f(x, y)$ is the integrated path traversed by the source as given by

$$f(x, y) = \int g(x, y, t)d(vt) \qquad (7.4)$$

where v is the source velocity in the direction of motion. The resultant recorded path, as given above, is

$$f\left(\frac{x_d}{k+m}, \frac{y_d}{k+m}\right).$$

This impulse response, because of the geometry, is independent of the lateral coordinates of the object and can thus be placed in convolutional form for each depth plane z. The resultant image due to a transparency t at plane z is given by

$$I_d(x_d, y_d) = t ** h$$

$$= t\left(\frac{x_d}{M}, \frac{y_d}{M}\right) ** Bf\left(\frac{x_d}{k+m}, \frac{y_d}{k+m}\right). \qquad (7.5)$$

In evaluating the normalizing constant B, we use similar reasoning to that employed in Chapter 4. We assume that the total number of photons used during the exposure is N. If the source is translated uniformly, the number of photons emitted per unit distance during the exposure is N/L, where L is the line integral of the path $f(x, y)$ as given by

$$L = \iint f(x, y)dxdy. \qquad (7.6)$$

Although this is a two-dimensional integration, it represents a line integral since $f(x, y)$ is a delta line function. Thus the constant B, representing the intensity per unit distance at the detector plane, is given by

$$B = \frac{KN}{L4\pi d^2(k+m)^2}\cos^3\theta = \frac{I_i}{L(k+m)^2} \qquad (7.7)$$

where I_i is the incident intensity as defined in (4.2), (4.4), and (4.5), and K is proportional to the energy per photon. Ignoring obliquity factors we can set $I_i = I_0$, the intensity at the axis, which assumes that $\cos^3\theta \simeq 1$. The resultant detected intensity becomes

$$I_d(x_d, y_d) = t\left(\frac{x_d}{M}, \frac{y_d}{M}\right) ** \frac{I_0}{L(k+m)^2} f\left(\frac{x_d}{k+m}, \frac{y_d}{k+m}\right). \quad (7.8)$$

The most widely used form of motion tomography is linear tomography, where the source and film are both moved uniformly in straight lines in opposite directions. The source motion is described by

$$g(x, y, t) = \delta(x - vt)\delta(y) \text{ rect}\left(\frac{vt}{X}\right) \quad (7.9)$$

where v is the velocity of the source in the x direction and X is the extent of the traverse. The resultant source path is given by

$$f(x, y) = \int g(x, y, t)d(vt) = \text{rect}\left(\frac{x}{X}\right)\delta(y) \quad (7.10)$$

indicating a line of length X in the x direction. The resultant detected image from (7.8) becomes

$$I_d(x_d, y_d) = t\left(\frac{x_d}{M}, \frac{y_d}{M}\right) ** \frac{I_0}{X(k+m)} \text{ rect}\left[\frac{x_d}{X(k+m)}\right]\delta(y_d). \quad (7.11)$$

The $\delta(y_d)$ in the expression above could be eliminated by defining the convolution as being one-dimensional in the x direction only. Thus each point at plane z is smeared into a horizontal line of length $X(k+m)$. At the desired plane z_0 we have

$$k = -m = \frac{d - z_0}{z_0} \quad (7.12)$$

where the rect function becomes a narrow delta function and reproduces the transparency in its original form. The plane of interest is at the depth $z_0 = d/(k+1)$. A chest tomogram using linear motion is shown in Fig. 7.2. Note the defocusing of the ribs and spine.

In the frequency domain the Fourier transform of t is multiplied by the Fourier transform of the point-spread function h. For the case shown above, we have

$$I_d(u, v) = I_0 M^2 T(Mu, Mv) \text{ sinc } [X(k+m)u]. \quad (7.13)$$

Thus, in the u direction, a plane at z is multiplied by a low-pass filter having an effective bandwidth of approximately $[X(k+m)]^{-1}$. At the desired plane $z = z_0$, this becomes an infinite-bandwidth filter and does not affect the frequency response $I_d(u, v)$. All other planes experience various degrees of filtering.

When using a finite source rather than a point source, the total impulse response is the convolution of the motion path function and source size as given by

$$h(x_d, y_d) = \frac{1}{4\pi d^2 m^2} s\left(\frac{x_d}{m}, \frac{y_d}{m}\right) ** \frac{1}{L(k+m)^2} f\left(\frac{x_d}{k+m}, \frac{y_d}{k+m}\right). \quad (7.14)$$

Thus at the tomographic plane $z = z_0$ where $m = -k$ and f becomes a delta function, the resolution continues to be limited by the source size. The expression in (7.14) reduces to that of the point source as given by (7.5) and (7.7) if we substitute $s(x, y) = KN\delta(x, y)$.

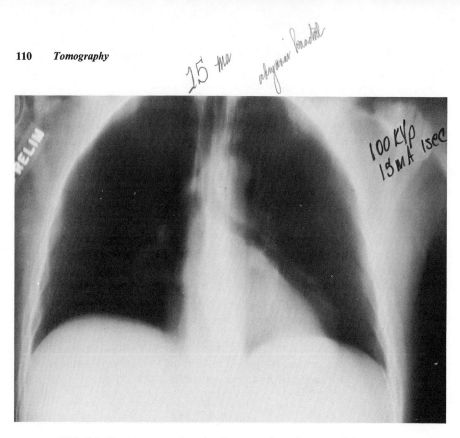

FIG. 7.2 Chest tomograph, using linear motion of source of detector.

Circular Motion

As was indicated, many other motions can be used other than the linear motion described. The linear motion has the disadvantage that planes other than the desired tomographic plane experience blurring in one dimension only. Thus edges parallel to the x axis receive no blurring at all. Structures of this type in any plane will remain sharply defined and can interfere with the visualization of structures in the desired plane. The alternative is the use of two-dimensional motions such as circles, hypercycloids, and so on. A circular motion can be described using the polar coordinate equivalent of the delta function where

$$\delta(x - x_0, y - y_0) = \frac{\delta(r - r_0)}{r}\delta(\theta - \theta_0). \qquad (7.15)$$

The circular motion is described by

$$g(x, y, t) = g(r, \theta, t) = \frac{\delta(r - r_0)}{r}\delta(\theta - \omega t)\,\text{rect}\left(\frac{\omega t}{2\pi}\right) \qquad (7.16)$$

representing a single circular traverse. The resultant source path is described as

$$f(x, y) = \int \frac{\delta(r - r_0)}{r} \delta(\theta - \omega t) \, \text{rect}\left(\frac{\omega t}{2\pi}\right) r d(\omega t)$$

$$= \delta(r - r_0) \, \text{rect}\left(\frac{\theta}{2\pi}\right) = \delta(r - r_0). \tag{7.17}$$

The film center also moves in a circle at the opposite side of the axis having a radius kr_0. When the source is at r_0, θ_0 the film center is at $kr_0, \theta_0 + \pi$. The resultant impulse response from (7.3) and (7.7) is given by

$$h(x_d, y_d) = h(r) = \frac{I_0}{2\pi r_0 (k + m)} \delta[r - (k + m)r_0]. \tag{7.18}$$

Motion tomography has two basic disadvantages. First, for each tomographic plane, the entire volume of interest is exposed by x-rays. If a number of sections are desired, as is usually the case, the radiation can be extensive. Second, in motion tomography, the detail contrast in the plane of interest is not improved over a conventional radiograph. All planes other than the plane of interest are blurred or smeared out, leaving the desired plane as the only one with any detailed structure. Thus the sharp details of the interfering structures in other planes are removed, which significantly improves the visualization even though the detail contrast in the desired plane is unchanged.

Multiple-Radiograph Tomography; Tomosynthesis

The second of the disadvantages, the detail contrast being the same as that of conventional radiography, is fundamental to motion tomography. The first of the disadvantages, however, can be remedied by a system shown in Fig. 7.3 known as *tomosynthesis* [Grant, 1972], where the desired plane is selected after the x-ray procedure. Here a sequence of different radiographs are taken with the source in different positions and the subject and film in the same posi-

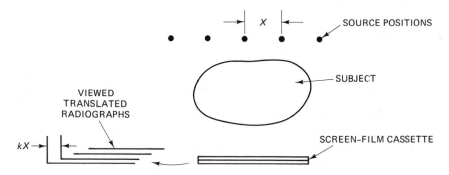

FIG. 7.3 Sequence of radiographs used to create a tomogram.

tion. The resultant films are stacked together and translated with respect to each other to select the desired plane. Assuming that the resultant composite image is the sum of the individual detected images, the path function $f(x, y)$ is a sum of point-source positions given by

$$f(x, y) = \delta(y) \sum_{i=-n/2}^{n/2} \delta(x - iX) \tag{7.19}$$

where n is the number of source positions and X is the separation between source positions. The value of k, representing the translation of the individual radiographs, is determined after the x-ray exposures are made, where each transparency is separated from its neighbor by kX. The resultant effective detected intensity for a given plane $t(x, y)$ using (7.8) is given by

$$I_d(x_d, y_d) = t\left(\frac{x_d}{M}, \frac{y_d}{M}\right) ** \delta(y_d)\frac{1}{n} \sum_{i=-n/2}^{n/2} \delta[x_d - i(k + m)X]. \tag{7.20}$$

The line integral L becomes n, the sum of the source positions, and the $(k + m)^2$ normalization in the denominator is canceled by the factors in the delta functions. It is clear that k is readily chosen to select the desired plane of interest. Since each film is given approximately $1/n$ of the exposure normally required, the system requires no increase in radiation dose for the ability to select planes after the exposure. The system requires a relatively rapid film changer so that the n exposures can be accomplished in a breath-holding interval of a few seconds. The out-of-focus planes are smeared by a series of points that approach a line. Using an appropriate mechanical structure the separation of the radiographs can be continuously varied with the plane of focus continuously moving through the object. Alternatively, the information can be collected and stored electronically, using television fluoroscopy, with the translation and summation taking place in a computer.

CODED SOURCE TOMOGRAPHY

Another approach to tomography is the use of a relatively large complex source $s(x, y)$. The recorded image, as studied in Chapter 4, is given by

$$s\left(\frac{x}{m}, \frac{y}{m}\right) ** t\left(\frac{x}{M}, \frac{y}{M}\right).$$

This recording can be considered the encoded image I_c. This encoded image is not useful of itself because of the complex source function.

The desired image, at any plane of interest, is decoded by cross correlation with $s(x/m_1, y/m_1)$, the source function at a particular plane z_1 where $m_1 =$

$-(d - z_1)/z_1$. The decoded image is given by

$$I_d = I_c \star\star s\left(\frac{x}{m_1}, \frac{y}{m_1}\right)$$

$$= \left[s\left(\frac{x}{m_1}, \frac{y}{m_1}\right) \star\star s\left(\frac{x}{m}, \frac{y}{m}\right)\right] \ast\ast t\left(\frac{x}{M}, \frac{y}{M}\right). \qquad (7.21)$$

The source function $s(x, y)$ is chosen to have a sharp autocorrelation peak, approaching a two-dimensional delta function. Thus at $z = z_1$, I_d will faithfully reproduce the plane $t(x/M_1, y/M_1)$ where $M_1 = d/z_1$. For $z \neq z_1$, the cross-correlation function will become broad, thus blurring all other planes. Some representative functions for $s(x, y)$ are a random array of points or a Fresnel zone plate. The considerations are similar to those of coded apertures in Chapter 8.

The decoding can be accomplished in a digital computer. For a given encoded image $I_c(x, y)$, any desired plane can be chosen by using the appropriate cross-correlation function $s(x/m_1, y/m_1)$. The basic difficulty with this system is that $s(x, y)$ is fundamentally a nonnegative function. It thus has limited capability for providing the desired autocorrelation peak. At best, the autocorrelation will have a peak riding on a large plateau. This plateau effectively represents an integration over a large portion of the image. This can distort the low-frequency response and result in poor noise performance. As a result, this approach has not been used commercially.

COMPUTERIZED TOMOGRAPHY

Motion tomography, at best, represents a limited ability to isolate a specific plane. In general the contrast of the plane of interest is unchanged over that of a projection radiograph. If a lesion in the plane results in a 1% difference in recorded intensity in a conventional radiograph, it will continue to be 1% different in the motion tomogram. The out-of-focus planes, however, will be blurred.

In 1973 a revolutionary concept in tomography, known as *computerized axial tomography*, was introduced by EMI Ltd. of England. This system provides an isolated image of a section within a volume completely eliminating all other planes [Herman, 1980; Gordon, 1975; Ledley, 1976; Scudder, 1978; Brooks and DiChiro, 1976a; Cho, 1974]. Thus the contrast of the image is not diminished by intervening structures. Thus far, computerized tomography has been extremely successful in clinical use. Lesions and organs that were heretofore impossible to visualize are seen with remarkable clarity.

The basic system is shown in Fig. 7.4. An x-ray source is collimated into a narrow beam and scanned through the plane of interest. The transmitted

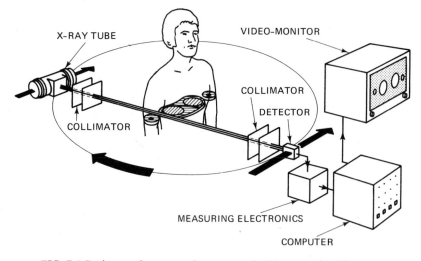

X-RAY TUBE

VIDEO-MONITOR

COLLIMATOR

DETECTOR

COLLIMATOR

MEASURING ELECTRONICS

COMPUTER

FIG. 7.4 Basic scanning system for computerized tomography. (Courtesy of the Siemens AG-Bereich Medizinische Technik.)

photons are detected by a scanning detector at each position of the scan. This same procedure is repeated at approximately 1° intervals for 180° so that a set of projections are obtained at approximately all angles. The resultant projection data are applied to a digital computer where an accurate two-dimensional image is reconstructed, representing the linear attenuation coefficient in the section of interest. The mathematics involved in the image reconstruction from projection data will be described.

This approach overcomes essentially all of the shortcomings of motion tomography. Only the section of interest is irradiated. Using carefully calibrated detectors, and limited only by the Poisson statistics of the number of counts per measurement, this technique has provided almost uncanny visualization of structures that were previously invisible. Radiologists have been able to perceive lesions whose attenuation coefficient differed by less than 0.5% from the surrounding tissue. Thus, in a noninvasive fashion, an accurate diagnosis is obtained.

RECONSTRUCTION MATHEMATICS— ITERATIVE APPROACHES

The mathematics involved is a relatively old, but seldom used, field of study involving the reconstruction of a two-dimensional distribution from its projections. The most straightforward, although computationally inefficient solution involves linear algebra. The two-dimensional image is reconstructed using a matrix inversion of the projection data. For images of reasonable complexity, this is quite formidable. One general class of solutions involves an iterative procedure. This is an attempt to find a two-dimensional distribution that matches all of the projections. An initial distribution is assumed and it is compared with the measured projections. Using one of a variety of iterative algorithms, the initial distribution is successively modified. This method is known as the *Algebraic Reconstruction Technique,* or ART [Herman, 1980; Brooks and DiChiro, 1976a].

The ART system, illustrated in Fig. 7.5, is based on the very general premise that the resultant reconstruction should match the measured projections. The iterative process is started with all reconstruction elements f_i set to a constant such as the mean $\bar{f}$ or zero. In each iteration the difference between

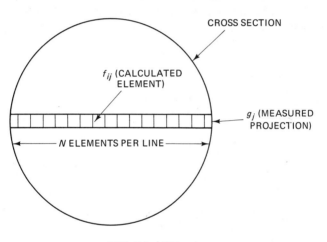

FIG. 7.5 ART system.

the measured data for a projection g_j and the sum of the reconstructed elements along that ray $\sum_{i=1}^{N} f_{ij}$ is calculated. Here f_{ij} represents an element along the jth line forming the projection ray g_j. This difference is then evenly divided among the N reconstruction elements. The iterative algorithm is defined as

$$f_{ij}^{q+1} = f_{ij}^{q} + \frac{g_j - \sum_{i=1}^{N} f_{ij}^{q}}{N} \tag{7.22}$$

where the superscript q indicates the iteration. The algorithm recursively relates the values of the elements to those of the previous iteration.

As an illustration of the ART process we use a simple 2×2 matrix of values and the associated measured projections.

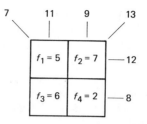

All six projection measurements, including the two verticals, two horizontals, and two diagonals, have been made. Presumably, these projection measurements are all that is available and, from these, the matrix of elements shown must be reconstructed. We begin the process arbitrarily by setting all values to zero, calculating the resultant projections, and comparing them to the measured projections. The differences are calculated, divided by the two elements per line, and added to each element.

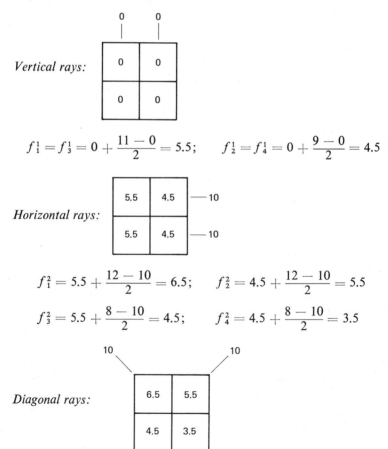

Vertical rays:

$$f_1^1 = f_3^1 = 0 + \frac{11 - 0}{2} = 5.5; \qquad f_2^1 = f_4^1 = 0 + \frac{9 - 0}{2} = 4.5$$

Horizontal rays:

$$f_1^2 = 5.5 + \frac{12 - 10}{2} = 6.5; \qquad f_2^2 = 4.5 + \frac{12 - 10}{2} = 5.5$$

$$f_3^2 = 5.5 + \frac{8 - 10}{2} = 4.5; \qquad f_4^2 = 4.5 + \frac{8 - 10}{2} = 3.5$$

Diagonal rays:

$$f_1^3 = 6.5 + \frac{7 - 10}{2} = 5; \quad f_2^3 = 5.5 + \frac{13 - 10}{2} = 7$$

$$f_3^3 = 4.5 + \frac{13 - 10}{2} = 6; \quad f_4^3 = 3.5 + \frac{7 - 10}{2} = 2$$

Thus the original elements are reconstructed. In general, for larger formats, many iterations, using the same measurement data over and over, are required for adequate convergence. The process is usually halted when the difference between the measured and calculated projections is adequately small.

A number of variations on this general theme have been proposed. One nonlinear formulation makes use of the known nonnegativity of the density values f_{ij}. Thus where $f_{ij} < 0$, it is set equal to zero. Another variation is known as *multiplicative ART*, as compared to the previous original algorithm, which is *additive ART*. In the multiplicative version the original density values are multiplied by the ratio of the measured line integral g_j to the calculated sum of the reconstructed elements. This is given by

$$f_{ij}^{q+1} = \frac{g_j}{\sum_{i=1}^{N} f_{ij}^q} f_{ij}^q. \tag{7.23}$$

In multiplicative ART, each reconstructed element is changed in proportion to its magnitude. This is in sharp contrast to additive ART, where each element in the ray is changed a fixed amount, independent of its magnitude.

Although the iterative methods were the most popular in the earlier days of computerized tomography, they have become almost completely supplanted by direct methods due to problems such as computation time and convergence accuracy in the presence of noise. The direct methods provide a linear reconstruction formulation between a two-dimensional distribution and its projections.

DIRECT RECONSTRUCTION METHODS— FOURIER TRANSFORM APPROACH

Direct reconstruction methods are based on the central section theorem, which is illustrated with the aid of Fig. 7.6. As shown, a single projection is taken in the x direction, for convenience, forming a projection $g(y)$ given by

$$g(y) = \int f(x, y)dx. \tag{7.24}$$

This projection represents an array of line integrals in the x direction. For demonstrating the central section theorem we use the two-dimensional Fourier

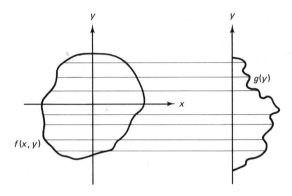

FIG. 7.6 Illustration of the central section theorem.

transform of the distribution $f(x, y)$ as given by

$$F(u, v) = \int\int f(x, y) \exp[-i2\pi(ux + vy)]dxdy. \qquad (7.25)$$

Along the $u = 0$ line this transform becomes

$$F(0, v) = \int\int f(x, y) \exp(-i2\pi vy)dxdy$$

$$= \int \left[\int f(x, y)dx \right] e^{-i2\pi vy} dy$$

$$= \mathfrak{F}_1\{g(y)\} \qquad (7.26)$$

where $\mathfrak{F}_1\{\cdot\}$ represents a one-dimensional Fourier transform. Thus, as shown in Fig. 7.7, the dashed $u = 0$ line in $F(u, v)$ is given by the Fourier transform

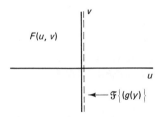

FIG. 7.7 Fourier domain illustration of the central section theorem.

of the projection of $f(x, y)$ in the x direction. Since the transform of each projection forms a radial line in $F(u, v)$, we can fill $F(u, v)$ by taking projections at many angles and taking their transforms. Once filled $F(u, v)$ is inverse transformed to reconstruct the desired density $f(x, y)$. This process can be studied in more detail using Fig. 7.8. Using a two-dimensional distribution $f(x, y)$, an array of line integrals are measured, each being a distance R from the origin

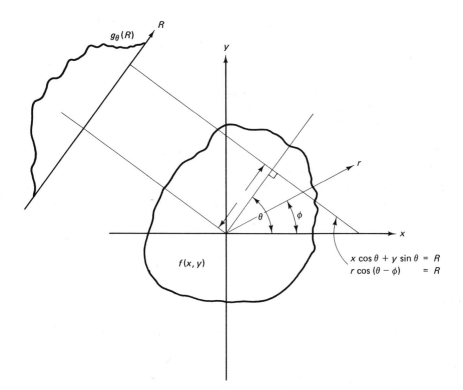

FIG. 7.8 Projection of a two-dimensional function.

where the perpendicular to the line is at an angle θ. This forms a projection given by

$$g_\theta(R) = \iint f(x, y)\delta(x \cos \theta + y \sin \theta - R)dxdy$$

$$= \int_0^{2\pi} \int_0^\infty f(r, \phi)\delta[r \cos (\theta - \phi) - R]rdrd\phi \qquad (7.27)$$

where $g_\theta(R)$ is the projection information in the θ direction. The integration takes place along line $x \cos \theta + y \sin \theta = R$ or, in polar coordinates r, ϕ, $r \cos (\theta - \phi) = R$. The delta line $\delta(x \cos \theta + y \sin \theta - R)$ sifts out the desired line in $f(x, y)$ to provide an effective line integration. The symbol $g_\theta(R)$ could alternatively have been written $g(R, \theta)$ since it is a two-dimensional function of the various projection angles θ and the distances R along each projection. However, the symbol $g_\theta(R)$ indicates a series of one-dimensional measurements at different distances R taken at a particular angle θ.

To provide a general derivation, the Fourier transform of the two-dimensional function $f(x, y)$ is given by

$$F(u, v) = \iint f(x, y)e^{-i2\pi(ux+vy)}dxdy. \qquad (7.28)$$

Expressing this in polar coordinates $F(u, v) = F(\rho, \beta)$, where $u = \rho \cos \beta$ and $v = \rho \sin \beta$ gives

$$F(\rho, \beta) = \iint f(x, y)e^{-i2\pi\rho(x \cos \beta + y \sin \beta)}dxdy. \tag{7.29}$$

The general central section theorem is shown by again mainpulating the two-dimensional Fourier transform to include the projection expression as given by

$$F(\rho, \beta) = \iiint f(x, y)\delta(x \cos \beta + y \sin \beta - R) \exp(-i2\pi\rho R)dxdydR. \tag{7.30}$$

This expression clearly reduces to the basic Fourier transform relationship of equation (7.28) by integrating over R. By isolating the expression for the projection from equation (7.27), the Fourier transform of the image $f(x, y)$ can be rewritten as

$$F(\rho, \beta) = \int g_\beta(R) \exp(-i2\pi\rho R)dR \tag{7.31}$$

Therefore,

$$F(\rho, \beta) = \mathscr{F}_1\{g_\beta(R)\}.$$

Thus the Fourier transform of a projection at angle β, as defined in Fig. 7.8, forms a line in the two-dimensional Fourier plane at this same angle. Since the projection angle θ and the resultant polar angle in the Fourier transform plane β are identical, we can use the same symbol θ for both. The transform of a projection in the transform plane, $F(\rho, \theta)$, is shown in Fig. 7.9. After filling the entire $F(\rho, \theta)$ plane with the transforms of the projections at all angles, the reconstructed density is provided by the two-dimensional inverse transform as

$$f(x, y) = \iint F(u, v) \exp[i2\pi(ux + vy)]dudv$$

$$= \int_0^{2\pi} d\theta \int_0^\infty F(\rho, \theta) \exp[i2\pi\rho(x \cos \theta + y \sin \theta)]\rho d\rho. \tag{7.32}$$

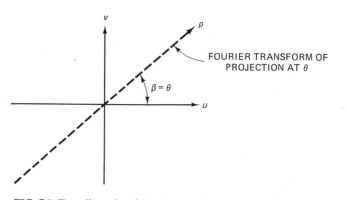

FOURIER TRANSFORM OF
PROJECTION AT θ

$\beta = \theta$

FIG. 7.9 Two-dimensional Fourier transform plane of distribution.

Figure 7.10 provides some physical insight into the central section theorem. In the top figure is a projection of a two-dimensional distribution $f(x, y)$. Using Fourier transform techniques $f(x, y)$ can be decomposed into an array of two-dimensional sinusoids. Two of the sinusoids are illustrated. In the center figure a projection is taken of a two-dimensional sinusoid. Since each ray experiences

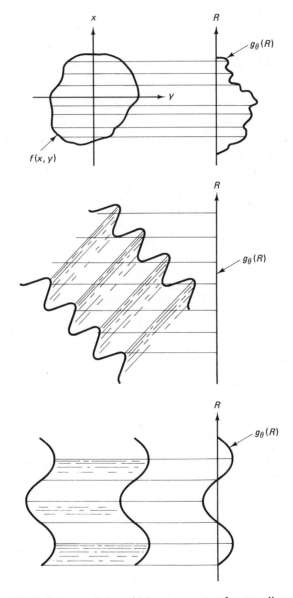

FIG. 7.10 Projections of sinusoidal components of a two-dimensional distribution.

equal positive and negative contributions, the resultant projection is zero. In the bottom figure the sinusoid is parallel to the projection angle. As a result, the projection is a one-dimensional sinusoid. Thus each projection extracts only those sinusoidal components at the projection angle. These components represent a line, as in Fig. 7.9, in Fourier space representing sinusoids of different frequencies at a specific angle.

EXAMPLES USING FOURIER TRANSFORM APPROACH

As an illustrative example we will consider two circularly symmetric cases of simple geometries. Circularly symmetric examples are being used solely for simplicity since the method is clearly applicable to any geometry. In the first example the measured projections at all angles are given by

$$g_\theta(R) = 2 \text{ sinc } 2R. \tag{7.33}$$

The two-dimensional Fourier transform at each angle θ is given by

$$F_\theta(\rho) = \mathscr{F}\{g_\theta(R)\} = \text{rect}\left(\frac{\rho}{2}\right), \tag{7.34}$$

where ρ, in this equation, is a one-dimensional variable along the θ direction. Adding up the contributions at all angles is equivalent to rotating the rect function over 180° to provide a symmetric two-dimensional frequency function

$$F(\rho, \theta) = F(\rho) = \text{circ } \rho, \tag{7.35}$$

where the circ function is a "pillbox" with unity radius and unity height as defined in Chapter 2. The reconstructed density is the inverse transform of this function as given by

$$f(x, y) = f(r) = \mathscr{F}^{-1}\{\text{circ } \rho\} = \frac{J_1(2\pi r)}{r}. \tag{7.36}$$

This "jinc" function, named for its similarity to the sinc function, has a similar shape to the sinc function except that its zeros do not occur periodically and the amplitude of the ripples fall off more rapidly. Thus the projection of $J_1(2\pi r)/r$ is 2 sinc 2R, a result that is certainly not intuitively obvious.

For the second example of circularly symmetric objects we use a cosinusoidal projection

$$g_\theta(R) = \cos \pi R. \tag{7.37}$$

As before, on each line in the transform, such as the u axis, we obtain

$$F(u) = \tfrac{1}{2}[\delta(u - \tfrac{1}{2}) + \delta(u + \tfrac{1}{2})].$$

Summing this pair of delta functions over 180°, we obtain

$$F(\rho) = \tfrac{1}{2}\delta(\rho - \tfrac{1}{2}) \tag{7.38}$$

representing a cylindrical shell in the frequency domain. The density is again given by the inverse transform. Using the Fourier–Bessel transform yields

$$f(r) = 2\pi \int_0^\infty F(\rho)\rho J_0(2\pi r\rho)d\rho = \frac{\pi}{2}J_0(\pi r). \tag{7.39}$$

Again we reach the unanticipated result that the projection of a $J_0(\cdot)$ function, the zeroth-order Bessel function of the first kind, is a cosine function.

ALTERNATIVE DIRECT RECONSTRUCTION— BACK PROJECTION

The computational problem with the central section theorem method shown is that a two-dimensional inverse transform is required. For computerized tomography this involves various interpolations and coordinate transformations. We now consider alternative reconstruction systems based on the same general principles but having distinct computational advantages. To do this we first introduce the concept of back projection [Gordon, 1975].

In back projection the measurements obtained at each projection are projected back along the same line, assigning the measured value at each point in the line. Thus the measured values are "smeared" across the unknown density function as if a line of wet ink, containing the measured projection values, is drawn across the reconstructed density function. This is shown in Fig. 7.11 for the case of an object consisting of a single point on the origin. Each projection is identical. Intuitively, we know, from each individual projection measurement, that a point of density lies somewhere along the line of

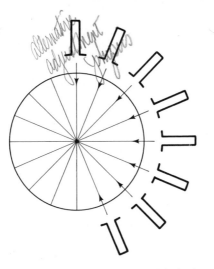

FIG. 7.11 Projections of a point at the origin back projected.

integration. It is thus reasonable, as an initial attempt at reconstruction, to assign the measured value along the entire line. We are essentially stating that we know that the point of density is somewhere along the line so that a "crude" reconstruction will result if we assign the measured value along the entire line.

Mathematically, the back projection of a single measured projection along the unknown density is given by

$$b_\theta(x, y) = \int g_\theta(R)\delta(x \cos \theta + y \sin \theta - R)dR \qquad (7.40)$$

where $b_\theta(x, y)$ is the back-projected density due to the projection $g_\theta(R)$ at angle θ. Adding up these densities at all angles, we obtain

$$
\begin{aligned}
f_b(x, y) &= \int_0^\pi b_\theta(x, y)d\theta \\
&= \int_0^\pi d\theta \int_{-\infty}^\infty g_\theta(R)\delta(x \cos \theta + y \sin \theta - R)dR \qquad (7.41)
\end{aligned}
$$

where $f_b(x, y)$ is the crude reconstruction resulting from pure back projection. This reconstruction is often called a *laminogram*. We will study the nature of the distortion in this reconstruction and attempt to correct it.

Using (7.27) and (7.41), we find the impulse response using back projection. We first find the projections $g_\theta(R)$ due to a delta function at the origin $\delta(r)/\pi r$ as given by

$$
\begin{aligned}
g_\theta(R) &= \int_0^{2\pi} \int_0^\infty \frac{\delta(r)}{\pi r}\delta[r \cos (\theta - \phi) - R]rdrd\phi \\
&= \int_0^\pi \int_{-\infty}^\infty \frac{\delta(r)}{\pi}\delta[r \cos (\theta - \phi) - R]drd\phi \\
&= \int_0^\pi \frac{\delta(R)}{\pi}d\phi = \delta(R).
\end{aligned}
\qquad (7.42)
$$

Thus, as expected intuitively, each projection of a delta function at the origin is $\delta(R)$. These delta functions are back-projected giving the impulse response $h_b(r)$:

$$
\begin{aligned}
h_b(r) &= \int_0^\pi d\theta \int_{-\infty}^\pi \delta(R)\delta[r \cos (\theta - \phi) - R] \, dR \\
&= \int_0^\pi \delta[r \cos (\theta - \phi)]d\phi \\
&= \int_0^\pi \frac{\delta\left[\theta - \left(\frac{\pi}{2} + \phi\right)\right]}{\left|\frac{\partial}{\partial\theta} r \cos (\theta - \phi)\right|}d\theta \quad \text{evaluated at } \theta = \frac{\pi}{2} + \phi \\
&= \frac{1}{r}
\end{aligned}
\qquad (7.43)
$$

where we have made use of the delta function relationship [Bracewell,1965]

$$\delta[f(x)] = \sum \frac{\delta(x - x_n)}{|f'(x_n)|} \tag{7.44}$$

where x_n are the roots of $f(x)$.

Knowing the impulse response to be $1/r$, we write the reconstructed image from back projection

$$f_b(x, y) = f(x, y) ** \frac{1}{r}. \tag{7.45}$$

This represents a poor reconstruction in the case of reasonably complex images because of the "tails" of the $1/r$ response. Some early reconstructions were obtained of medical images using pure back projection with marginal results.

The $1/r$ blurring must be removed to provide the desired reconstruction. One approach makes use of the frequency-domain representation of (7.45), where

$$F_b(\rho, \theta) = \frac{F(\rho, \theta)}{\rho} \tag{7.46}$$

since the two-dimensional Fourier transform of $1/r$ is $1/\rho$. An obvious correction method is to take the Fourier transform of $f_b(x, y)$, weight the resultant $F_b(\rho, \theta)$ with ρ, and then inverse transform to provide the desired $f(x, y)$. This, however, has clearly not solved the computational problem since two two-dimensional transforms are required.

FILTERED BACK-PROJECTION RECONSTRUCTION SYSTEM

It is desirable to be able to use the elegant simplicty of back projection and to undo the $1/r$ blur without requiring two-dimensional transforms. This is accomplished by again making use of the powerful central section theorem.

We begin by restating the back-projection relationship for the laminogram $f_b(x, y)$ as

$$f_b(x, y) = \int_0^\pi d\theta \int_{-\infty}^\pi g(R, \theta)\delta(x \cos \theta + y \sin \theta - R)dR \tag{7.47}$$

and restructuring it into a Fourier transform mode by using the central section theorem to substitute the inverse transform of $F(\rho, \theta)$ for $g(R, \theta)$ as given by

$$f_b(x, y) = \int_0^\pi d\theta \int_{-\infty}^\infty \left[\int_{-\infty}^\infty F(\rho, \theta)e^{i2\pi\rho R}d\rho \right] \delta(x \cos \theta + y \sin \theta - R)dR. \tag{7.48}$$

Performing the integration over R, we obtain

$$f_b(x, y) = \int_0^\pi d\theta \int_{-\infty}^\infty F(\rho, \theta) \exp [i2\pi\rho(x \cos \theta + y \sin \theta)]d\rho. \tag{7.49}$$

To appreciate the significance of this relationship we restate the two-dimensional inverse Fourier transform relationship in polar form as

$$f(x, y) = \int_0^{2\pi} d\theta \int_0^\infty F(\rho, \theta) \exp\left[i2\pi\rho(x \cos\theta + y \sin\theta)\right]\rho \, d\rho. \quad (7.50)$$

We modify this equation to conform with (7.49), where the θ integration is from 0 to π and the ρ integration from $-\infty$ to ∞ as given by

$$f(x, y) = \int_0^\pi d\theta \int_{-\infty}^\infty F(\rho, \theta) \exp\left[i2\pi\rho(x \cos\theta + y \sin\theta)\right]|\rho| \, d\rho. \quad (7.51)$$

This form is equivalent to that of (7.50), where the integrand is Hermitian. This is clearly the case for physical systems where $f(x, y)$ is real and $F(\rho, \theta) = F(-\rho, \theta + \pi)$. It is necessary to use $|\rho|$ rather than ρ in (7.51) since the integration includes negative values.

Comparing (7.51) to the equation for the back-projected laminogram (7.49), we see that they differ only by the $|\rho|$ weighting as previously indicated. Substituting $\mathcal{F}_1\{g_\theta(R)\}$ for $F(\rho, \theta)$ in (7.49), and dividing and multiplying by $|\rho|$, we obtain

$$f_b(x, y) = \int_0^\pi d\theta \int_{-\infty}^\infty \frac{\mathcal{F}_1\{g_\theta(R)\}}{|\rho|} \exp\left[i2\pi\rho(x \cos\theta + y \sin\theta)\right]|\rho| \, d\rho. \quad (7.52a)$$

This equation provides an alternative interpretation to back projection. In essence, the transform of each projection $g_\theta(R)$ has been weighted by $1/|\rho|$ along each radial line in Fig. 7.9. This accounts for the blurred reconstruction of $f_b(x, y)$. This can therefore be removed by weighting each transformed projection with $|\rho|$ prior to back projection to create an undistorted reconstruction as given by

$$f(x, y) = \int_0^\pi d\theta \int_{-\infty}^\infty \frac{\mathcal{F}_1\{g_\theta(R)\} \cdot |\rho|}{|\rho|} \exp\left[i2\pi\rho(x \cos\theta + y \sin\theta)\right]|\rho| \, d\rho. \quad (7.52b)$$

When we operate on each projection $g_\theta(R)$ the radial frequency variable ρ assumes the role of a one-dimensional frequency variable.

This reconstruction approach is known as the filtered back projection system. It is widely used since it involves only one-dimensional transforms. Each projection is individually transformed, weighted with the one-dimensional variable $|\rho|$, inverse transformed, and back projected. This is seen by rewriting (7.52) as

$$f(x, y) = \int_0^\pi d\theta \int_{-\infty}^\infty \left[\int_{-\infty}^\infty \mathcal{F}_1\{g_\theta(R)\} \cdot |\rho| e^{i2\pi\rho R}\right] \delta(x \cos\theta + y \sin\theta - R) dR \quad (7.53)$$

$$= \int_0^\pi d\theta \int_{-\infty}^\infty \mathcal{F}_1^{-1}[\mathcal{F}_1\{g_\theta(R)\} \cdot |\rho|] \delta(x \cos\theta + y \sin\theta - R) dR. \quad (7.54)$$

Here it is clearly seen that the function back-projected at all angles is a filtered version of the projection $g_\theta(R)$, where the filter provides a $|\rho|$ weighting. As in any filtering operation in the frequency domain, we first Fourier transform, multiply by the filter function, and then inverse transform.

$$\int_0^h \int_{-\infty}^\infty g_\theta(R)\, \delta(x\cos\theta + y\sin\theta - R)\, dR\, d\theta$$

$$= \int_0^\pi \int$$

CONVOLUTION—BACK PROJECTION

The back-projected function in (7.54) can be rewritten as

$$\mathcal{F}_1^{-1}[\mathcal{F}_1\{g_\theta(R)\}\cdot|\rho|] = g_\theta(R) * \mathcal{F}_1^{-1}\{|\rho|\} \qquad (7.55)$$

using the convolution theorem of Fourier transforms. The spatial equivalent of filtering with $|\rho|$ is convolving with the inverse transform of $|\rho|$ [Horn, 1978; Tanaka, 1979; Scudder, 1978]. This introduces the convolution–back projection method of reconstruction, which is by far the most widely practiced. Instead of filtering each projection in the frequency domain, each projection $g_\theta(R)$ is convolved with a function $c(R)$ and then back-projected. Since the convolution function $c(R)$ is chosen to correct the $1/r$ blur, the reconstruction is exact as given by

$$f(x,y) = \int_0^\pi d\theta \int_{-\infty}^\infty [g_\theta(R) * c(R)]\delta(x\cos\theta + y\sin\theta - R)dR. \qquad (7.56)$$

As indicated in (7.55), the convolution function is given by

$$c(R) = \mathcal{F}^{-1}\{|\rho|\}. \qquad (7.57)$$

Unfortunately, this transform is not defined since the function is not integrable. However, we can evaluate the transform in the limit as

$$c(R) = \mathcal{F}^{-1}\left\{\lim_{\epsilon\to 0} |\rho|e^{-\epsilon|\rho|}\right\} \qquad (7.58)$$

which is an integrable function. Evaluating the transform, we have

$$|\rho|e^{-\epsilon|\rho|} = \rho[e^{-\epsilon\rho}H(\rho) - e^{+\epsilon\rho}H(-\rho)] \qquad (7.59)$$

where $H(\cdot)$ is the unit step function, which is unity for positive arguments and zero otherwise, as defined in Chapter 2. We first find the inverse transform of the bracketed portion as given by

$$\mathcal{F}^{-1}\{[\cdot]\} = \int_0^\infty e^{-\epsilon\rho}e^{i2\pi R\rho}d\rho - \int_{-\infty}^0 e^{\epsilon\rho}e^{i2\pi R\rho}d\rho$$

$$= \frac{i4\pi R}{\epsilon^2 + (2\pi R)^2}. \qquad (7.60)$$

The completed inverse transform is evaluated using the relationship

$$\mathcal{F}^{-1}\{\rho A(\rho)\} = \frac{1}{2\pi}a'(R). \qquad (7.61)$$

Therefore,

$$\mathcal{F}^{-1}\{|\rho|e^{-\epsilon|\rho|}\} = \mathcal{F}^{-1}\{\rho[\cdot]\} = \frac{2(\epsilon^2 - 4\pi^2 R^2)}{(\epsilon^2 + 4\pi^2 R^2)^2}. \qquad (7.62)$$

The convolution function $c(R)$ is therefore the limit of (7.62) as $\epsilon \to 0$.

The various properties of this convolution function can be put in a logical framework, keeping in mind the fact that the convolution function is the inverse transform of $|\rho|$. Using known theorems of Fourier transforms, we know that the average value of a function is equal to the value of its transform at the

127

origin. Thus, since $|\rho|$ is zero at the origin, we know that the average integrated value of $c(R)$ must be zero. This is the case for the function given by (7.62). Using this same relationship in reverse, we know that the value of $c(R)$ at the origin must approach the integrated average of $|\rho|$. The area of $|\rho|$ is equal to the limit, as $\rho \longrightarrow \infty$, of $\frac{1}{2}\rho^2$. This is in keeping with the value of $c(R)$ at the origin in the limit, as $\epsilon \longrightarrow 0$, of $2/\epsilon^2$.

Figure 7.12 illustrates a simplified sketch of the convolution function $c(R)$. In the vicinity of the origin, where $R^2 \ll \epsilon^2/4\pi^2$, the function is given by $2/\epsilon^2$. For higher values of R, where $R^2 \gg \epsilon^2/4\pi^2$, the function approaches $-1/2\pi^2R^2$, as shown in Fig. 7.12.

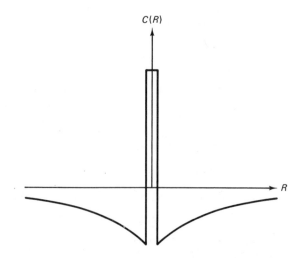

C(R)

R

FIG. 7.12 Convolution function to undo $1/r$ blur.

It is interesting to note that the desired convolution function $c(R)$ could have been devised without using the central section theorem. We can make use of linear systems considerations in that any function $c(R)$ which results in the proper reconstruction of an impulse, without the $1/r$ blur, will accurately reconstruct all functions. For example, in equation (7.56), we require a $c(R)$ that will produce an impulsian $f(x, y)$ when $g_\theta(R) = \delta(R)$, the projection of an impulse at the origin. Essentially, this requires that $c(R)$, back-projected at all angles, will produce an impulsian reconstruction.

For the $c(R)$ in the limit given in (7.62), back-projecting at all angles provides the system impulse response as given by

$$h(r) = \int_0^\pi d\theta \int_{-\infty}^\infty c(R)\delta[r\cos(\theta - \phi) - R]dR$$

$$= \lim_{\epsilon \to 0} 2 \int_0^\pi \frac{\epsilon^2 - 4\pi^2r^2\cos^2(\theta - \phi)}{[\epsilon^2 + 4\pi^2r^2\cos^2(\theta - \phi)]^2}d\theta. \tag{7.63}$$

Performing this complex integration, we arrive at the result

$$h(r) = \lim_{\epsilon \to 0} K \frac{\epsilon}{(\epsilon^2 + 4\pi r^2)^{3/2}} \qquad (7.64)$$

where K is a constant. This is clearly an "impulsian" function having a value of K/ϵ^2 at the origin and zero elsewhere. Other forms of $c(R)$ in the limit provide similar results [Horn, 1978].

We now consider more realistic convolution functions which have well-behaved properties. For example, any physical system has an upper frequency limit imposed by either the geometry of the system, such as a finite beam size, or by electrical limitations such as noise. Thus the filtering imposed by the convolution filter could be of the form $|\rho| \operatorname{rect}(\rho/2\rho_0)$, where ρ_0 is the cutoff frequency. This filter is illustrated in Fig. 7.13.

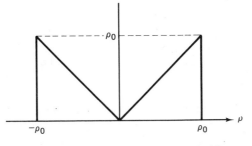

FIG. 7.13 Filtering function with band limiting.

To facilitate taking Fourier transforms, this function is modeled as

$$|\rho| \operatorname{rect}\left(\frac{\rho}{2\rho_0}\right) = \left[\operatorname{rect}\left(\frac{\rho}{2\rho_0}\right) - \Lambda\left(\frac{\rho}{\rho_0}\right)\right] \qquad (7.65)$$

where the convolution function is given by

$$\mathfrak{F}^{-1}\left\{|\rho| \operatorname{rect}\left(\frac{\rho}{2\rho_0}\right)\right\} = \rho_0(2 \operatorname{sinc} 2\rho_0 R - \operatorname{sinc}^2 \rho_0 R). \qquad (7.66)$$

Thus a bandlimited function can be reconstructed by taking each projection $g_\theta(R)$, convolving with $\rho_0(2 \operatorname{sinc} 2\rho_0 R - \operatorname{sinc}^2 \rho_0 R)$, and back projecting. If we take the limit of this bandlimited function as $\rho_0 \to \infty$, we get an expression similar to that of (7.62).

In practice, a wide variety of convolution functions are used which are similar to that of (7.65). The convolution kernel $c(R)$ is a convenient place to accomplish the overall system filtering in addition to the basic $|\rho|$ filtering. In general, as with most imaging systems, these filters are a compromise between signal-to-noise ratio and resolution. Thus these filters generally involve the product of $|\rho|$ and some high-frequency cutoff filter. Since the sharp cutoff of a rect function generally causes "ringing" at edges, gentler rolloff characteristics are usually used.

RECONSTRUCTION OF THE ATTENUATION COEFFICIENT

The reconstruction methods studied indicated how a two-dimensional function could be reconstructed from its line integrals. These line integrals are the sum of the function in different directions. In the case of x-ray attenuation, however, we measure the exponent of the desired line integral as given by

$$I = I_0 e^{-\int \mu(z) dz} \tag{7.67}$$

where, for convenience, we consider a single ray in the z direction. Note that we have a nonlinear relationship between the measured projection and the desired line integral. This non linearity can be removed if we use, as our measured data, the log of the measured transmission where

$$\ln\left(\frac{I_0}{I}\right) = \int \mu(z)\, dz. \tag{7.68}$$

In x-ray attenuation considerations, the logarithm of the measured intensity at many positions and angles is used with one of the reconstruction methods previously described to form a cross-sectional image of the linear attenuation coefficients.

The removal of the nonlinearity has within it a number of inherent assumptions, namely, that the source is monoenergetic and that the beam is infinitesimally narrow. Unfortunately, both of these assumptions lead to relatively impractical systems as far as getting sufficient photon flux to obtain a statistically meaningful measurement. Thus, to provide a source of adequate strength, a polychromatic source is used providing a measured transmission given by

$$I = \int S(\mathcal{E}) \exp\left[-\int \mu(z, \mathcal{E}) dz\right] d\mathcal{E} \tag{7.69}$$

where $S(\mathcal{E})$ is the source spectrum. In this case the line integral is not directly measured and the resultant behavior is nonlinear [Stonestrom et al., 1981]. Here we attempt to reconstruct $\mu(\bar{\mathcal{E}})$, where $\bar{\mathcal{E}}$ is the average energy emerging from the object. Taking logs as in (7.68), we obtain

$$\ln\left(\frac{I_0}{I}\right) = a_0 + a_1 \int \mu(z, \bar{\mathcal{E}})\, dz + a_2 \left[\int \mu(z, \bar{\mathcal{E}}) dz\right]^2 + \ldots \tag{7.70}$$

where $I_0 = \int S(\mathcal{E}) d(\mathcal{E})$, the total source energy. We have a distorted version of the desired line integral.

The nature of the distortion can be seen by studying a single pixel in a cross section traversed by a number of rays at different angles. Along each ray we are attempting to measure the line integral or the sum of the attenuation coefficients of each pixel. The attenuation coefficient of the single pixel being studied should contribute a given μ to each sum. However, each ray can contain different materials providing different degrees of spectral shift and a resultant different average energy $\bar{\mathcal{E}}$. Thus the single pixel has different contributions to each ray sum, resulting in a distortion in the attempted reconstruction. From this discussion it is seen that the largest distortions will occur in the vicinity of bone where the greatest spectral shifts occur. In early head scans this effect resulted in a severely cusped region immediately inside the skull, making diagnosis of this region very difficult.

These nonlinear terms cause distortions in the reconstructed image which can be quite severe. As a result, most instruments use a nonlinear function of the log of the measurements in an attempt to compensate for the nonlinearity. Unfortunately, the degree of nonlinearity depends on the materials in the path, which are not known beforehand. This problem has been minimized in some computerized tomography scanners by using a water bag around the region being scanned, thus providing a constant known path length. This approach still provides some error since the amount of bone and air within each path is still unknown. Other approaches are under investigation. In one, an initial reconstruction is provided which includes the distortions or spectral shift artifacts. From this initial reconstruction the amount of bone and soft tissue in each path can be estimated and used to provide a more accurate nonlinear correction.

The other nonlinearity is that due to finite beam size. The projection for a finite beam size is given by

$$I = \iint s(x, y) \exp\left[-\int \mu(x, y, z)dz \right]dxdy \qquad (7.71)$$

where $s(x, y)$ is the source intensity as a function of its lateral dimensions. For clarity we assume a monoenergetic beam in this case. As before, the error with this system depends on the variation in $\mu(x, y)$ over the beam size $s(x, y)$. If, over the beam region, $\mu(x, y)$ is relatively constant at some value $\bar{\mu}$, the measurement can be approximated as

$$\ln\left(\frac{I_0}{I}\right) \simeq \int \bar{\mu}(z)dz \qquad (7.72)$$

where

$$I_0 = \iint s(x, y)dxdy.$$

Unfortunately, there are many discontinuities in the attenuation coefficient, such as the edges of bone, so that this approximation is often inaccurate. This remains a source of error in existing instruments which is minimized through the use of relatively narrow beams.

SCANNING MODALITIES

In Fig. 7.4 we illustrated a simple method of data acquisition where a single source and detector are synchronously scanned to provide the required projection data. This system, because of chronology, is known as a "first-generation" scanner. It is identified by a two-motion translate–rotate scan using a single detector. The principal difficulty with this instrument is its relatively long scan time, on the order of a few minutes. Only a small portion of the total x-ray output of the x-ray tube is utilized, requiring relatively long scanning times to achieve adequate statistics. These long times are acceptable, however, for relatively stationary regions such as the head. These scanners continue to achieve relatively wide use because of their low cost. However, even as a head scanner, there are continuing problems with uncooperative patients such as children and patients with poor motor control.

Figure 7.14 is an illustration of a second-generation scanner at various intervals of the scan. Here the same translate–rotate motions are used with a multiple-detector system. In this way, several projections are acquired during each traverse. For example, if there are 10 detectors, each 1° apart, a single translation acquires all 10 projections. During the subsequent rotation the gantry is indexed 10° rather than 1°, resulting in a 10:1 time reduction. Since 10 times as much of the x-ray output is being utilized, the scan time is cut accordingly. Using this approach, scan times have been reduced to a fraction of a minute.

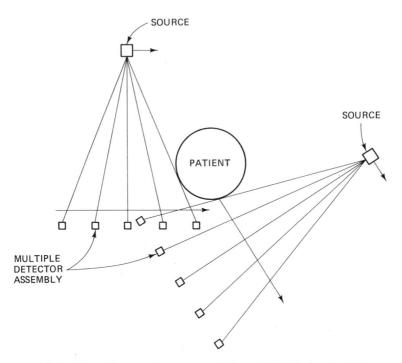

FIG. 7.14 Second-generation scanner, using multiple detector translate-rotate system.

One important feature of both first- and second-generation scanners is self-calibration. Either preceding or following each traverse, each x-ray beam impinges on the detector with no intervening material. This provides a reference measurement of I_0, the intensity in the absence of attenuation. This value is required to calculate the line integral. Although it is theoretically constant, drifts in both source and detector often require frequent measurements.

The third-generation scanner involves rotation-only of a fan beam, as illustrated in Fig. 7.15. Both the source and the detector are rotated about a common center within the patient. The detector array is a few hundred contiguous elements. The primary advantage of this approach is the mechanical simplicity and associated ability to provide very high speeds, with scan times as low as 1 sec. The detectors can be relatively deep and positioned along the rays radiating from the source. This relatively long path length has enabled the use of gaseous ionization detectors using xenon.

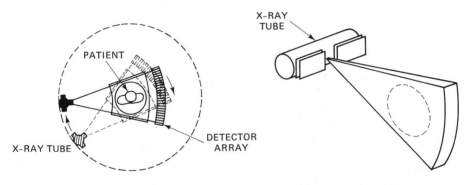

FIG. 7.15 Third-generation fan-beam scanner, using a rotating source and detector array. (Courtesy of the General Electric Medical Systems Division.)

One disadvantage of this system is the lack of self-calibration. At no point after the patient enters the machine can the system be calibrated. In the earlier days of these instruments "ring artifacts" were prevalent due to errors in individual detectors which were uncalibrated. These have since been minimized through improved detectors and software corrections. A commercial third-generation system is illustrated in Fig. 7.16.

The fourth-generation scanner is characterized by a rotating fan beam impinging on a 360° stationary detector array as illustrated in Fig. 7.17. A source, generating a fan beam, is rotated around the patient. The transmitted rays are collected by the stationary detector array. This simple mechanical motion of the source only allows for a rapid scan time. In addition, the system is again self-calibrating since, at different portions of the scan, each detector is irradiated by the source without any intervening material. Also, the system is relatively

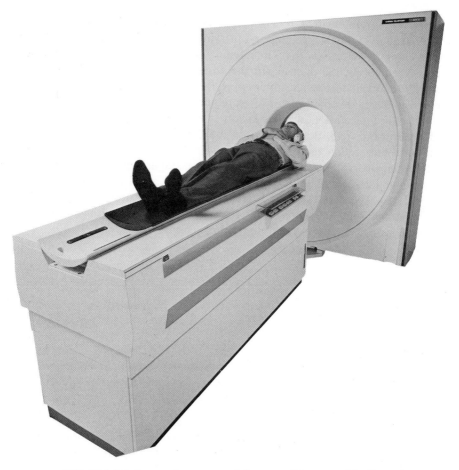

FIG. 7.16 Third-generation commercial scanner. (Courtesy of the General Electric Medical Systems Division.)

immune to ring artifacts since detector errors are distributed throughout the image, rather than representing a specific radius.

One difficulty with the scanner is the varying angle at which the rays strike the detectors. In the third-generation scanner the detectors could be aligned along the rays since the entire structure rotated. Here, however, at different source positions the rays strike a given detector at different angles. This means that the detectors should be relatively shallow to avoid the rays entering adjacent detectors. To provide high quantum efficiency with these shallow detectors, high-μ materials are used such as scintillators with high atomic numbers. Gaseous detectors, having lower linear attenuation coefficients, are not used with stationary detector systems.

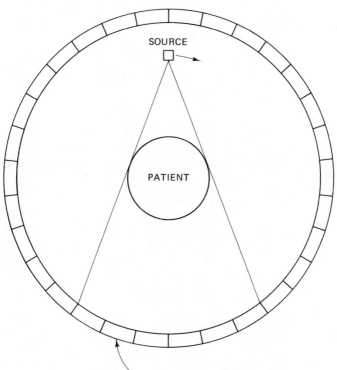

FIG. 7.17 Fourth-generation scanner, using a rotating fan-beam and a stationary ring detector array.

The third- and fourth-generation scanners derive their measurements using fan-beam rather than parallel-beam projections. These require somewhat modified approaches to the reconstruction problem. One approach is known as re-binning, where the various fan-beam rays from different projections are reassembled as parallel-beam projections. These then require the same reconstruction algorithms as the first- and second-generation scanners. An alternative approach is the use of a modified convolution back-projection system [Gullberg, 1979; Denton et al., 1979]. Here the convolution kernel is slightly different and the back projection involves a quadratic weighting factor rather than the uniform weighting of the parallel rays. This latter algorithm is widely used in existing scanners.

The image quality of these systems has improved significantly in recent years, as has their diagnostic value. Typical head and body images made with a third-generation scanner are illustrated in Figs. 7.18 and 7.19.

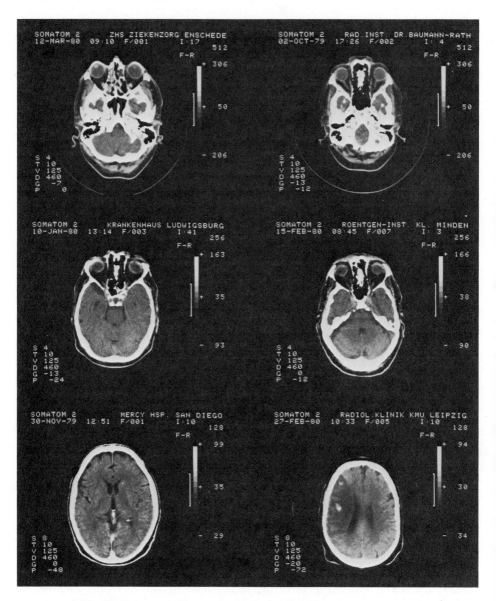

FIG. 7.18 Cross-sectional head images at different levels, made with a third-generation scanner. (Courtesy of the Siemens AG-Bereich Medizinische Technik.)

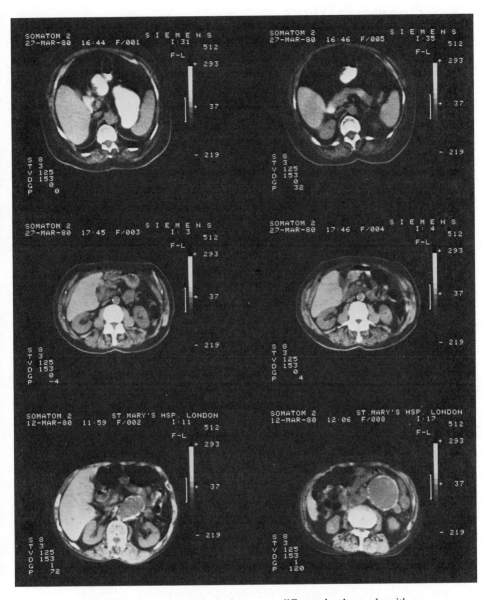

FIG. 7.19 Cross-sectional body images at different levels, made with a third-generation scanner. (Courtesy of the Siemens AG-Bereich Medizinische Technik.)

NOISE CONSIDERATIONS
IN COMPUTERIZED TOMOGRAPHY

The noise considerations of the measurements in computerized tomography are similar to those of projection radiography [Chesler et al., 1977; Brooks and DiChiro, 1976a]. We have an array of independent measurements, each having the general form

$$N_i = n_0 A \exp\left(-\int_i \mu dl\right)$$ (7.73)

where N_i is the number of detected photons at each measurement, n_0 is the incoming photon density in photons per unit area, A is the active area of the detector receiving the impinging x-ray beam, and $\int_i \mu dl$ is the ith line integral of the cross section $\mu(x, y)$, representing one of the rays. For simplicity, the collection efficiency η is assumed to be unity, a reasonable assumption with available detectors used in CT. As in projection radiography, the noise or standard deviation of each measurement is $\sqrt{N_i}$ because of the Poisson statistics.

In computerized tomography, however, we first calculate the line integrals of μ using logs and then, using the line integrals, reconstruct the values of μ. The line integrals are given by

$$g_i = \int_i \mu dl = \ln\left(\frac{N_0}{N_i}\right)$$ (7.74)

where $N_0 = n_0 A$, the incident number of photons per measurement. Using this relationship, we calculate the mean $\bar{g}_i$ and σ_{g_i}, the standard deviation of the line integral resulting from the uncertainty in the measurement of N_i.

The mean and variance of the line integral of the projection have been derived in Chapter 6. For a reasonably large number of photons per measurement N_i, the mean and variance of the line integral are given by

$$\bar{g}_i \simeq \ln\left(\frac{N_0}{\bar{N}_i}\right)$$ (7.75)

$$\sigma_{g_i}^2 \simeq \frac{1}{\bar{N}_i}$$ (7.76)

where $\bar{N}_i$ is the mean of the number of counts per measurement.

Using these statistics of the line integral measurement of the reconstructed image $\mu(x, y)$, we wish to analyze the signal-to-noise ratio as given by

$$\text{SNR} = \frac{C\bar{\mu}}{\sigma_\mu}$$ (7.77)

where C, as with projection systems, is the fractional change in μ, $\bar{\mu}$ is the mean, and σ_μ is the standard deviation. We now proceed to calculate $\bar{\mu}$ and σ_μ for a set of discrete projections in an appropriately normalized fashion. For convenience we introduce the notation $R' = r \cos (\theta - \phi)$ so that the back-projection operator takes the form $\delta(R - R')$.

Using the convolution–back projection system, the resultant reconstruction is given by

$$\mu(x, y) = \int_0^\pi [g_\theta(R') * c(R')]d\theta \qquad (7.78)$$

where $g_\theta(R)$ represents the projection at the angle θ, $c(R)$ is the convolution function, and each has been convolved with $\delta(R - R')$, the back-projection operator. For studying noise we use a realistic model of M discrete projections as given by

$$\hat{\mu} = \sum_{i=1}^M [g_{\theta_i}(R') * c(R')]\Delta\theta \qquad (7.79)$$

where $\hat{\mu}$ is the estimate of μ. To evaluate the required normalization and facilitate the use of transforms, we express this finite sum in integral form as

$$\hat{\mu} = \frac{M}{\pi} \int_0^\pi [g_\theta(R') * c(R')]d\theta \qquad (7.80)$$

where $\Delta\theta = \pi/M$. This can be expressed as a two-dimensional convolution of the desired function $\mu(r, \phi)$ as

$$\hat{\mu} = h(r, \phi) ** \mu(r, \phi) \qquad (7.81)$$

where $h(r, \phi)$ is the two-dimensional impulse response as given by

$$h(r, \phi) = \frac{M}{\pi} \int_0^\pi c[r \cos (\theta - \phi)]d\theta. \qquad (7.82)$$

For this analysis it is convenient to normalize the area of $h(r, \phi)$ to unity so that the reproduced values of $\hat{\mu}$, in broad constant areas, will not be scaled and will represent the correct average value. We thus require that

$$\int_0^{2\pi} \int_0^\infty h(r, \phi)r\,dr\,d\phi = 1. \qquad (7.83)$$

The impulse response having a unity area is equivalent to its Fourier transform $H(\rho)$ being unity at the origin.

Taking transforms using equation (7.82), we obtain

$$H(\rho) = \frac{M}{\pi} \frac{C(\rho)}{|\rho|} \qquad (7.84)$$

where $H(\rho)$ is the transform of the circularly symmetric impulse response and $C(\rho)$ is the Fourier transform of the convolution function $c(R)$. Division by $|\rho|$ is again the affect of back projection. $C(\rho)$ can be decomposed as given by

$$C(\rho) = |\rho| S(\rho) \qquad (7.85)$$

where $|\rho|$ removes the back-projection blur and $S(\rho)$ is the system filter. Thus the normalization procedure requires that

$$S(0) = \frac{\pi}{M}.$$

For example, using the rectangular filter of Fig. 7.13, we obtain

$$C(\rho) = |\rho| \frac{\pi}{M} \operatorname{rect}\left(\frac{\rho}{2\rho_0}\right). \tag{7.86}$$

We will calculate the signal-to-noise ratio using this filter function.

In taking the variance of the estimate of μ we use the statistical property that the variance of a sum of independent measurements is equal to the sum of the variances. For a weighted sum of measurements, the weightings are squared in accordance with the definition of the variance in (2.49). Since each measurement is weighted by $c(R)$, the resultant measurement variance is weighted by $[c(R)]^2$. The variance of the reconstruction is therefore the variance of the measurements convolved with $[c(R)]^2$ and back-projected. In integral form this is given by

$$\sigma_\mu^2 = \frac{M}{\pi} \int_0^\pi \sigma_{g_\theta}^2(R') * [c(R')]^2 d\theta \tag{7.87}$$

where $\sigma_{g_\theta}^2(R)$ is the variance of g_θ along each projection. In the continuous case, from (7.76), it is given by

$$\sigma_{g_\theta}^2(R) = \frac{1}{\bar{n}_\theta(R)h} \tag{7.88}$$

where $\bar{n}_\theta(R)$ is the average transmitted photon density and h is the height of the beam normal to the section. The $\bar{n}h$ represents the number of transmitted photons per unit distance along the projection.

To achieve useful results, we assume a typical radiographic object where the density of transmitted photons $\bar{n}_\theta(R)$ is relatively constant and can be approximated by a constant $\bar{n}$. In that case the convolution becomes an integral of $c^2(R)$ and the variance becomes

$$\sigma_\mu^2 = \frac{M}{\pi} \frac{1}{\bar{n}h} \int_0^\pi d\theta \int_{-\infty}^\infty c^2(R)dR = \frac{M}{\bar{n}h} \int_{-\infty}^\infty c^2(R)dR. \tag{7.89}$$

We evaluate this integral using Parseval's theorem, giving

$$\sigma_\mu^2 = \frac{M}{\bar{n}h} \int_{-\infty}^\infty |C(\rho)|^2 d\rho. \tag{7.90}$$

Using the normalized rectangular filter from (7.86), we obtain

$$\sigma_\mu^2 = \frac{M}{\bar{n}h} \int_{-\infty}^\infty \rho^2 \frac{\pi^2}{M^2} \operatorname{rect}\left(\frac{\rho}{2\rho_0}\right) d\rho = \frac{\pi^2}{\bar{n}hM} \frac{2\rho_0^3}{3}. \tag{7.91}$$

The resultant signal-to-noise ratio is therefore

$$\text{SNR} = \frac{C\bar{\mu}}{\sigma_\mu} = \frac{C\bar{\mu}\sqrt{\frac{3}{2}\bar{n}hM}}{\pi}\,\rho_0^{-3/2}. \tag{7.92}$$

It is more useful to structure the result, as before, in terms of resolution limitations imposed by the detector width w so that the trade-off between resolution and signal-to-noise ratio is readily visualized. Obviously, increasing the bandwidth ρ_0 without bound would be poor design since the signal-to-noise ratio would become poorer while the resolution would continue to be limited by the effective detector width w. Thus, in a good design, the bandwidth ρ_0 is compatible with a resolution equal to the detector width. Thus we have

$$\rho_0 = \frac{K}{w} \tag{7.93}$$

where K is a constant of order unity depending on the shape of the system response. This constant will vary slightly depending on which resolution criteria is used. The signal-to-noise ratio then becomes

$$\text{SNR} = K'C\bar{\mu}\sqrt{\bar{n}hM}\,w^{3/2} \tag{7.94}$$

where K' is a combined constant, again of order unity.

It is indeed interesting to study the implications of the resultant signal-to-noise ratio. Considerable insight can be derived by structuring this relationship in terms of $\bar{N}$, the average number of counts per measurement as given by

$$\text{SNR} = K'C\bar{\mu}\sqrt{\bar{N}M}\,w \tag{7.95}$$

where $\bar{N} = \bar{n}A = \bar{n}wh$.

In our noise studies in projection radiography in Chapter 6, the signal-to-noise ratio was shown to be dependent solely on the number of counts per measurement. The counts per measurement were governed by the imput radiation, the body attenuation, and the area of a pixel. Here we see the additional factor w. Thus, in computerized tomography, a higher-resolution system suffers an additional noise penalty, over and above the reduced number of photons per element for a given dose. This increased penalty is a result of the convolution operation, which couples the noise values of adjacent measurements into each measurement. The procedure that decouples the signal information, removing the $1/r$ blur, increases the noise since the signal is subjected to a $c(R)$ convolution while the noise variance experiences a $c^2(R)$ convolution as indicated in (7.87).

The analysis using a continuous measured projection $g_\theta(R)$ ignored the finite width of the detector w. Effectively, for a simple single scanned detector system as shown in Fig. 7.4, the projection is being convolved with a detector or beamwidth function such as rect (R/w). This can be treated as part of the overall convolution function $c(R)$. Thus $c(R)$ becomes the convoution of rect (R/w) and the function used for reconstruction. For detector arrays this situation becomes more complex because of the aliasing, which is introduced when $g_\theta(R)$ is sampled.

$$\frac{1}{4\pi d^2 u^2}\, \delta\!\left(\frac{x_d}{m}, \frac{y_d}{m}\right) * * \qquad \frac{1}{X(k+u)^2}$$

PROBLEMS

7.1 In a linear tomography system a source a distance d from the film is moved uniformly an amount X in the x direction with the film moved an amount kX in the opposite direction. A sinusoidal transparency having a transmission

$$t = a + b \cos 2\pi f_0 x$$

is imaged. At what depths z will the sinusoidal component at f_0 disappear?

7.2 In an x-ray imaging system the desired information is at plane $z = d/2$ and the undesired structure, at plane $z = 2d/3$, consists of a symmetrical square wave in the x direction of period W.
(a) Find the parameters k and X of a linear motion tomography system that will focus on the desired plane and eliminate the square-wave structures.
(b) At what other depth planes will this square-wave grating disappear?

7.3 An x-ray source, parallel to and a distance d from the x-ray recorder, has a pattern in the x direction of rect (x_0/X). It is moved linearly in the x direction an amount D with the recorder moved in the opposite direction an amount kD where $D > X$.
(a) Find the two z distances for placing a transparency at which the point-spread function due to the source alone is equal to that due to the motion alone.
(b) Plot the point response at the recorder, in the x direction, for these two z planes labeling the break points.
(c) Plot the point response for planes at $z = d, z = d/(1 + k), z = d/(1 + k/2)$.

7.4 In a linear tomography system a source is moved a distance A in the x direction with the recorder moved kA in the opposite direction. The source distribution in the x direction is rect (x/X), a distance d from the recorder.
(a) At what two depth planes is the response a rect function? What are the widths?
(b) At what two planes is the response a triangular function, and what is the width of the response at each plane?

7.5 In a linear tomography system the source is translated in the x direction at a velocity v for a time interval τ. The recorder is linearly translated in the opposite direction with a velocity kv for the same time interval. The source distribution in the x direction is given by rect (x/X).
(a) Find the overall point response in the x direction as a function of z.
(b) Find the thickness of the tomographic cut, which is defined as the

$$\frac{1}{4\pi d^2 u^2}\, \theta \; \text{rect}\!\left(\frac{x}{uX}\right)\delta(y) * * \frac{l_0}{L(k+u)^2}\,\text{rect}\!\left(\frac{x}{kHu}\right)\!\delta(y)$$

$$A\!\left(k+u\right) = uuX$$

distance between those z planes where the size of the response due to the source motion alone is equal to the response size due to the source alone.

7.6 An x-ray imaging system consists of a circular disk source of radius r_s separated a distance d from a film–screen system which has an impulse response circ (r/r_f). The source is linearly translated in the x direction an amount A with the film moved kA in the opposite direction.
(a) Find the impulse response of the system for a transparency at plane z.
(b) A transparency consists of two pinholes separated in the x direction by S. Over what depth range can the transparency be placed with the resultant images separated, that is, not overlapping?
(c) Repeat part (b) for two holes of radius r_h.

Make the necessary assumptions concerning the relative dimensions such that the images are separable in the absence of translation.

7.7 (a) Find the projection space $g(R, \theta)$ of a two-dimensional function $f(x, y) = \cos 2\pi f_0 x$. Using the filtered back-projection reconstruction system, find the back-projected function. Show that this function results in a correct reconstruction.
(b) Repeat part (a) for $f(x, y) = \cos 2\pi ax + \cos 2\pi by$ and $f(x, y) = \cos 2\pi(ax + by)$.

7.8 Find the circularly symmetric function $f(r)$ which has a projection at all angles of

$$p(R) = \sqrt{1 - R^2} \, \text{rect} \frac{R}{2}.$$

[*Hint:* Use Fourier transform tables in Bracewell (1965).]

7.9 The area of a two-dimensional function is $\iint f(x, y)dxdy$.
(a) Find an expression for the area in terms of the projection $g_\theta(R)$.
(b) Show that the function $h_R(R)h_\theta(\theta)$ cannot represent a projection $g(r, \theta)$ unless $h_\theta(\theta)$ is a constant.

7.10 In a computerized tomography system each projection is obtained using a uniform scanning beam of width W instead of an infinitesimal pencil beam. Find the resultant estimate $\hat{f}(x, y)$ of the function $f(x, y)$ using a conventional reconstruction system that does not take the beam width into account.

7.11 Find the signal-to-noise ratio of the computerized tomography reconstruction of a lesion immersed in a 20-cm cylinder of water whose attenuation coefficient is 5% different than that of the water. A scanned source is used providing 100 projections at 0.1R per projection. The detector and beam dimensions are 2.0 × 2.0 mm. Make appropriate assumptions about the reconstruction filter.

7.12 Projections $g_\theta(R)$ are taken of a unit square where $f(x, y) = \text{rect } (x) \text{ rect}$ (y).

(a) Find a general expression for $g_\theta(R)$ and the particular functions for $\theta = 0°$ and $45°$.

(b) Using the method of filtered back projection, find the *Fourier transform* of the back-projected function for the general case and for $\theta = 0°$ and $45°$.

7.13 In the convolution back projection reconstruction system, find a general expression for the impulse response of the reconstruction $h(r)$ for a convolution function $c(R)$. The answer can be left in integral form and should be a function of r and R only.

$$\text{sinc}\left(\rho \cos \theta\right)$$
$$\text{sinc}\left(\rho \sin \theta\right)$$

7.12 .

$$g_\theta(R) \Rightarrow f(x, y) = \text{rect } (x) \text{ rect } (y)$$

rect

$$\int g_\theta(R) \, \delta($$

$$g_\theta$$

$$\int_\theta \theta$$

$$\tan \theta$$

$$x \cos \theta + y \sin \theta =$$

$$\frac{1}{2} \sin \theta + x \cos \theta$$

$$x \cos \theta + y \sin \theta = \rho$$

$$x \leq \frac{1}{2}, \, \frac{1}{2}$$
$$y \leq \frac{1}{2}, \, \frac{1}{2}$$

$$\int f(x, y)$$

$$\int \delta(x \cos \theta$$

$$\int_{\frac{1}{2}}^{?} \int_{?}^{\frac{1}{2}} \delta x$$

$$-\frac{1}{2} \quad \frac{1}{2}$$

$$\frac{\text{sinc}\left(\rho\cos\theta\right)}{\sin\theta\cos\theta}\,\text{sinc}\left(\frac{\rho\sin\theta}{\cos\theta}\right)$$
$$\Rightarrow \text{rect}\left(\frac{R}{\cos\theta}\right)*\text{rect}\left(\frac{R}{\sin\theta}\right)$$

8

$$\cos\theta + \sin\theta \cdot \frac{1}{\left(\dfrac{0}{\cos\theta}\right)}$$

Nuclear Medicine

$$\frac{\sin\theta}{} \cdot \text{rect}\left(\frac{R}{\sin\theta}\right)$$

In radiography the regions under study are used in a transmission mode in the measurement of the attenuation coefficient. This measurement is often enhanced by the selective administration of radiopaque contrast materials. In nuclear medicine [Blahd, 1965] the region under study becomes an active source. This is done through the selective administration of radioactive materials since the body contains no natural radioactive substances. Either the radioactive material itself, or the chemical form it is bound in, has properties that cause it to be selectively taken up in specific regions of the anatomy. Once taken up in the organs of interest, these become radiating sources. Thus the imaging problem in nuclear medicine is that of defining a three-dimensional source distribution rather than a distribution of attenuation coefficients.

It is important to point out that, in general, much smaller amounts of administered materials are required in nuclear medicine than in radiographic contrast studies. The radiation dose problems are also quite different. In radiography the patient is irradiated only during the time the x-ray beam is turned on. In nuclear medicine the patient is being irradiated from the moment the radioactive material is administered until it is either eliminated by the body or decays.

The earliest nuclear medicine studies were made on the thyroid gland by taking advantage of its natural affinity for iodine. An isotope of iodine ^{131}I was used as the tracer material. This emits gamma rays at an energy of 364 kev.

Gamma rays are photons having the same energy range as x-rays. The definition simply distinguishes their source, with x-rays generated by electron events and gamma rays generated by nuclear events. The energy of 364 kev is somewhat high by radiography standards. In nuclear medicine, however, attenuation of the photons is undesirable. The requirement for negligible attenuation would of itself suggest very high energy isotopes. However, this requirement is moderated by the imaging considerations. Obtaining good collimation and efficient detection is difficult at very high energies, so that the energies used are often a compromise between the attenuation and imaging requirements.

In recent years ^{99m}Tc, an isotope of technicium [Blahd, 1965], has gotten wide acceptance as the preferred material for a variety of studies. This decision is based primarily on three desired properties. First, it is relatively easily made by chemical generators rather than requiring a cyclotron. Second, it has a gamma-ray emission energy of 140 kev, which is a good compromise between body attenuation and imaging considerations. Third, it has a relatively short half-life of 6 hours, for low radiation dosage.

SCANNED DETECTORS

In the early thyroid studies the imaging device was a simple single-bore, lead collimator which was scanned over the area of interest as shown in Fig. 8.1. Photons passing through the hole gave up their energy in a scintillating crystal whose material is comparable to that used in x-ray screens. Materials such as sodium iodide are used with the resultant visible photons coupled to a photomultiplier tube where an electrical signal is created. The crystal is made relatively thick so as to capture the gamma rays with almost 100% quantum efficiency. The number of light photons produced per gamma-ray photon and the light quantum efficiency of the photocathode are sufficiently high so that the signal-to-noise ratio is dominated by the number of received gamma-ray photons. The detector is mechanically scanned across the area of interest. The resultant signal is used to intensity modulate a synchronously scanned display which produces an image of the radioactivity distribution. Some scanners use

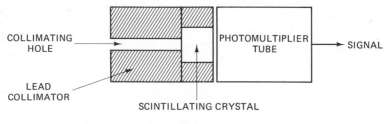

FIG. 8.1 Scanned gamma-ray detector.

focused collimators, aimed at a specific depth range, to provide a limited degree of depth resolution.

The relative radioactivity, often specified as representing hot and cold regions, is of diagnostic significance. In the thyroid these indicate regions of overactivity and underactivity. In other organs lesions can be demonstrated by regions of either underactivity or overactivity as compared to the normal organ. For example, brain tumors tend to localize the radioactive material and produce a hot spot. Liver studies, on the other hand, are produced by injecting a colloidal radioactive substance which is taken up in normal liver tissue. Thus regions of disease, such as tumors, are characterized by cold spots which do not take up the administered colloidal material.

Scanned systems are unsuitable for studying the dynamics of the functioning of an organ or system because of the long time required to view the area of interest. To accomplish this function cameras have been developed which view an entire region at once and make a series of images that indicate the distribution of the radioactive material at different time intervals. The ability to make these images rapidly is governed by the efficiency of the camera and the amount of radioactive material used. This amount is limited by radiation considerations.

The basic unit of activity is the curie (Ci), which is defined as 3.7×10^{10} disintegrating nuclei per second. In the case of a gamma-ray emitter this represents the number of photons emitted per second. The amount of radiation dose is thus determined by the amount administered, the half-life of the material, and the ability of the body to excrete the material. Iodine 131 is used in thyroid studies and has a half-life of 8.1 days. A dose of 25 μCi results in a radiation dose of 40 rad, a number significantly greater than any radiographic study. In the early days, using relatively inefficient detectors, doses of 50 to 250 μCi were used. With present-day detectors, doses of 5 μCi and less are used.

Scanned detector systems have rapidly declined in popularity, although they continue to be used for visualizing static structures of the body. This is particularly true of studies involving very large fields of view, such as a whole-body bone scan [Laughlin et al., 1960]. Figure 8.2 illustrates a commercial scanner together with a typical whole-body bone scan. As can be seen, the image is relatively noisy and of considerably lower resolution than typical radiographic images.

IMAGING CONSIDERATIONS
WITH GAMMA-RAY CAMERAS

The gamma-ray camera allows an entire field to be studied simultaneously without requiring a mechanical scan. For both static and dynamic studies, these are rapidly becoming the most widely used instruments. A basic gamma-ray camera is shown in Fig. 8.3.

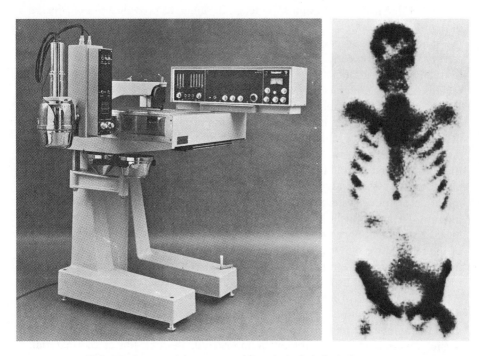

FIG. 8.2 Commercial scanner and a typical whole-body bone scan.

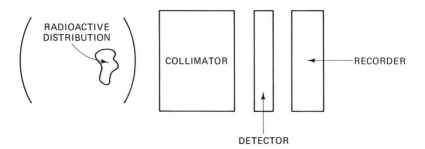

FIG. 8.3 Basic gamma-ray camera.

The camera consists of a collimator for forming the radioactive distribution into a two-dimensional image, a detector for detecting the position of each gamma-ray photon, and a recorder for producing an image from the detected photons. The detector considerations are similar to those of radiography. The material should be sufficiently thick and of high attenuation coefficient so as to stop most of the high-energy photons and produce large numbers of visible photons. A typical nuclear medicine detector has a $\frac{1}{2}$-inch-thick sodium iodide crystal.

The recorder must register the position of each event and form an image.

Film, the recorder used in radiography, is unsuitable because of scatter consid- erations. A large number of the emitted photons are Compton scattered and appear in the detector at erroneous positions. In radiography these are atten- uated by collimating grids since we know which direction the desired rays are coming from. In nuclear medicine, the direction of the desired rays is unknown. Scatter discrimination is provided by making use of the nature of Compton scattering described in Chapter 3. Each scattered photon has a reduced photon energy. At the relatively high energies used in nuclear medicine isotopes, this energy change is significant enough to be measured. Also, the photon rate is sufficiently slow that individual photons can be distinguished. Each single photon captured in the detector produces a number of visible photons propor- tional to the energy of the gamma ray. Desired events can be distinguished from scattered events by the amplitude of the light pulse produced. Thus electronic recorders, such as an array of photomultipliers with pulse-height analyzers, can minimize the scattered radiation. This capability is unavailable using photo- graphic film. In addition, having the signal in an electronic form makes possible a variety of processing, such as geometric distortion correction.

One generalized configuration for a detector and recorder utilizes an array of individual detectors. The output from each detector represents the integrated number of photons over a given incremental area. The detected outputs can be coupled to a recorder, where an image is produced of the intensity at the detector plane. Arrays of this form have been used employing scintillating crystals fol- lowed by photomultipliers [Bender and Blau, 1963]. In recent years suitable two- dimensional arrays have been built using gaseous multiwire proportional counters and also using cooled semiconductor arrays such as intrinsic and lithium-drifted silicon and germanium. In the detector arrays considered, the system resolution is limited to the number of discrete detectors used. With some detectors, such as sodium iodide scintillating crystals followed by photomulti- pliers, this can be a very awkward and expensive configuration if reasonable resolution is desired.

ANGER CAMERA

The Anger camera, named after its inventor, is a system for achieving a large number of resolvable elements with a limited number of detectors [Anger, 1958]. It thus overcomes the previous difficulty of having the resolution limited by the number of discrete detectors. The principle is based on estimating the position of a single event by measuring its contribution to a number of detectors. This system requires that the detector be capable of distinguishing individual events, no matter where they occur. Two simultaneous events occurring at different portions of the detector system would be rejected by this camera, whereas it could be recorded by the previously described array of individual detectors.

The basic principle is simply illustrated in Fig. 8.4 with a single slab of scintillating crystal followed by two photocells centered at x_1 and x_2. The light received from each detector, due to single events, is distributed among the two detectors based on the position of the event. Thus the position of the event $\bar{x}$ can be estimated as

$$\bar{x} \simeq \frac{I_1 x_1 + I_2 x_2}{I_1 + I_2}. \tag{8.1}$$

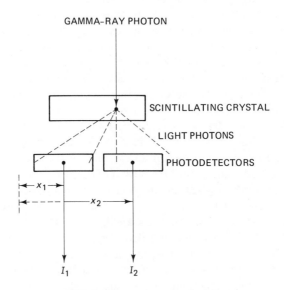

FIG. 8.4 Anger camera principle.

The ultimate resolution in the x direction, using only two detectors, is determined by the accuracy of the formula and, more important, the statistics of each measurement rather than by the size or number of detector elements. It must be emphasized, however, that the process requires detecting single events. If more than one scintillation takes place during the measurement interval, the position cannot be determined. Fortunately, the $I_1 + I_2$ sum can be used to indicate the sum of the counts received so as to discard multiple events and scatter. Cameras of this general type have a single crystal viewed by arrays of detectors with the detected outputs followed by a position computer to estimate the position of each event.

A typical detector system of this type is shown in Fig. 8.5. The light from the crystal is divided among the photomultiplier tubes arranged in a hexagonal array. The sum of the outputs is used for energy selection to achieve scatter rejection. If the pulse height fails to fall within prescribed limits for the isotope used, the pulse is rejected by blanking the intensity of the display device. This same process also rejects occasional multiple events. The position of the event

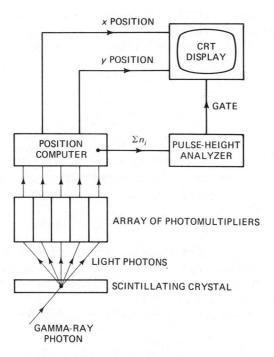

x POSITION

y POSITION

CRT DISPLAY

GATE

POSITION COMPUTER

Σn_i

PULSE-HEIGHT ANALYZER

ARRAY OF PHOTOMULTIPLIERS

LIGHT PHOTONS

SCINTILLATING CRYSTAL

GAMMA-RAY PHOTON

FIG. 8.5 Block diagram of an Anger camera.

on the crystal is determined by a centroid calculation. The estimate of the x and y coordinates of the event, $\hat{x}$ and $\hat{y}$, are given by

$$\hat{x} = \frac{\sum_i x_i n_i}{\sum n_i} \tag{8.2}$$

and

$$\hat{y} = \frac{\sum_j y_j n_j}{\sum n_j} \tag{8.3}$$

where x_i and y_j are the x and y coordinates of the center of the photomultipliers and n_i and n_j are the number of light photon counts or the pulse amplitudes in each photomultiplier. This system provides a resolution of over 1000 resolvable elements using 19 photodetectors. This is made possible by analyzing single events. However, even the 1000 elements represent a lateral resolution of about 1 cm, considerably less than that used in radiography.

The resolution of the camera, rather than depending on the number of discrete detectors, is determined by the accuracy of the position computation. This is limited by the counting statistics of the number of light photons at each photomultiplier. In calculating the statistics of the position measurement we make the simplifying assumption that the total number of collected counts $\sum_i n_i$ is a constant equal to N. First, with a large number of counts, the statistical variation in this quantity will be small compared to that of each n_i. Second, the pulse-height analysis will reject events whose total count is not in the immediate vicinity of N. The expected value of $\hat{x}$ is therefore given by

$$E(\hat{x}) = \frac{1}{N} \sum_i x_i E(n_i)$$

$$= \sum_i x_i \frac{\Omega_i}{\sum_j \Omega_j} \tag{8.4}$$

where Ω_i is the solid angle subtended by the ith photomultiplier from the light emitted from the scintillating crystal at point x, y and $\Omega_i / \sum_j \Omega_j$ is the fraction of the total collected counts in the ith photomultiplier. The solid angle at each photomultiplier due to an event at x, y can be approximated as

$$\Omega_i \simeq \frac{Ad}{[(x_i - x)^2 + (y_i - y)^2 + d^2]^{3/2}} \tag{8.5}$$

where A is the area of the photomultiplier cathode and d is the z distance from the event to the photocathodes. The distance d is assumed constant since the thickness of the crystal is smaller than the distance from the crystal to the photomultipliers as shown in Fig. 8.5.

The variance of the measurement, which determines the accuracy, is given by

$$\operatorname{var} \hat{x} = \sum_i \left(\frac{x_i}{N}\right)^2 \operatorname{var} n_i$$

$$= \sum_i \frac{x_i^2}{N} \frac{\Omega_i}{\sum_i \Omega_i} . \tag{8.6}$$

The standard deviation of the computation is therefore

$$\sigma_{\hat{x}} = \left\{ \frac{1}{N} \frac{\sum_i x_i^2 [(x_i - x)^2 + (y_i - y)^2 + d^2]^{-3/2}}{\sum_j [(x_j - x)^2 + (y_j - y)^2 + d^2]^{-3/2}} \right\}^{1/2} \tag{8.7}$$

where the area of the photomultiplier A cancels out. This area determines the overall collection efficiency which governs N, the total number of counts, as given by

$$N = \eta_p \frac{\lambda_{\text{light}}}{\lambda_{\text{x-ray}}} \frac{\sum_i \Omega_i}{4\pi} \tag{8.8}$$

where η_p is the efficiency of light production of the scintillating material. The calculations are essentially identical for $\sigma_{\hat{y}}$.

In Table 8.1 we list the standard deviation for a number of cases. For simplicity, a square $L \times L$ array of contiguous square photodetectors has been used. The spatial resolution or number of elements along each axis can be approximated by $L/2\sigma_{\hat{x}}$, with the total number of resolvable elements being the square of that number. A total number of 1000 collected photons is assumed for N. As shown in Table 8.1, the accuracy increases as the number of photomul-

TABLE 8.1

PERFORMANCE OF ANGER CAMERA

x, y Position of Source (cm)	Array Size, *L* (cm)	Photo-multiplier Size (cm)	Total Number of Photo-multipliers	Detector to Scintillator Distance *d* (cm)	Standard Deviation (cm)
0, 0	40	4	100	5	0.225
0, 0	40	8	25	5	0.25
0, 0	40	8	25	3	0.224
0, 0	40	4	100	3	0.177
20, 20	40	4	100	3	0.578
20, 20	40	8	25	3	0.614
20, 20	40	8	25	5	0.578
10, 10	40	8	25	3	0.368
0, 0	20	4	25	3	0.131
5, 5	20	4	25	3	0.186
10, 10	20	4	25	3	0.28

tipliers increase. The accuracy for points on the axis is improved by decreasing the distance d from the photodetectors to the scintillator. However, for off-axis sources the accuracy deteriorates at a more rapid rate for the smaller d. Also, a smaller field of view provides improved accuracy at the price of viewing a limited portion of the anatomy.

It should be emphasized that the centroid calculation of the position given in equations (8.2) and (8.3) are not optimum from a statistical point of view. The uniform weighting given each measurement can be shown to be suboptimal. Newer forms of position arithmetic [Gray and Macovski, 1976] using nonuniform weightings of the measurements are being used to provide reduced standard deviations for a given photon count.

PINHOLE IMAGING STRUCTURES

The imaging portion or collimator of the camera system in Fig. 8.3 contributes significantly to the determination of the efficiency and the lateral resolution. Radiographic systems do not need an imaging structure since the transmitted photons define the image through shadowing. In nuclear medicine we are imaging the source and thus require an imaging or collimating structure. Lenses are unavailable at this energy range since the refractive index of all transparent materials is unity. Since only attenuation mechanisms are available, a pinhole becomes the logical imaging device. This is known in nuclear medicine as a pinhole collimator and is shown in Fig. 8.6.

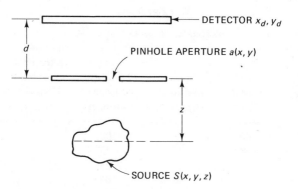

FIG. 8.6 Pinhole imaging system.

If the pinhole were an infinitesimally small opening on the axis of the system, a planar source at z, $S(x, y)$, would be reproduced at the detector plane as

$$I_d(x_d, y_d) = KS\left(\frac{x_d}{M}, \frac{y_d}{M}\right)$$ (8.9)

where the magnification M is given by

$$M = -\frac{d}{z}.$$ (8.10)

Although this system would have perfect fidelity in lateral resolution, it would have zero intensity since K, which is determined by the solid angle of interception, would approach zero.

To provide adequate photon flux the pinhole is opened to an aperture $a(x, y)$. We make the assumption that the aperture plate is infinitesimally thin and is completely opaque to the emitted gamma rays. Although this combination of assumptions is somewhat inconsistent, it does make the system space invariant and allows us to place the detected output in a convenient convolutional form. An aperture plate of finite thickness will have a space-variant point response because the sides of the aperture will alter its effective size when viewed from different angles.

To calculate the impulse response we place a single photon point source at x, y, z. The intensity or photon density at the detector plane due to photons entering the aperture, using the same development as given in equations (4.2) to (4.5), is $[1/4\pi(z + d)^2] \cos^3 \theta$, where θ is the angle the rays make with the normal. The extent of this intensity pattern is the magnified projected aperture function projected from the point source. This magnified image of the aperture is translated by Mx and My, respectively, where M is the image magnification $-d/z$. Thus the impulse response at the detector due to a point source is given by

$$h(x, y, x_d, y_d) = \frac{\cos^3 \theta}{4\pi(z + d)^2} \, a\left(\frac{x_d - Mx}{m}, \frac{y_d - My}{m}\right)$$ (8.11)

where m is the magnification of the projected aperture as given by

$$m = \frac{z + d}{z} = 1 - M. \tag{8.12}$$

In the interest of providing a space-invariant impulse response, with a subtle loss in accuracy, we ignore the $\cos^3 \theta$ dependence of the oblique rays. Since the system is linear we use the superposition integral to find the output $I_d(x_d, y_d)$ due to a general planar source $S(x, y)$ at plane z as given by

$$I_d(x_d, y_d) = \frac{1}{4\pi(z + d)^2} \int\int S(x, y) a \left(\frac{x_d - Mx}{m}, \frac{y_d - My}{m} \right) dx\,dy. \tag{8.13}$$

The previous approximation, which neglected the finite thickness of the pinhole aperture, and the $\cos^3 \theta$ obliquity factor have enabled us to structure this expression in convolutional form. We do this by making the substitutions

$$x' = Mx \quad \text{and} \quad y' = My$$

giving

$$I_d(x_d, y_d) = \frac{1}{4\pi d^2 m^2} S \left(\frac{x_d}{M}, \frac{y_d}{M} \right) ** a \left(\frac{x_d}{m}, \frac{y_d}{m} \right). \tag{8.14}$$

The capture efficiency of the system is simply based on the solid collection angle Ω. Using the same simplifying approximations used in developing the impulse response, we have

$$\eta(z) = \frac{\Omega}{4\pi} \simeq \frac{\int\int a(x, y)\,dx\,dy}{4\pi z^2} = \frac{A_p}{4\pi z^2}. \tag{8.15}$$

Here we see the basic trade-off between efficiency and resolution as represented by the area A_p of the pinhole.

It is instructive to examine equation (8.14) to evaluate the intensity of the image as a function of depth. If the source is a small point, the intensity of the image will decrease with increasing depth as indicated by the $(z + d)^2$ factor in the denominator of (8.11). Thus, as the point source is moved farther away, the projected aperture image will become both smaller and less intense.

The situation is different, however, for a large extended source whose lateral extent is appreciably greater than the extent of the projected aperture. Although the resolution and the magnification of the image vary with depth, the resultant intensity is depth independent. This can be appreciated by noting that the area of $a(x/m, y/m)/m^2$ is equal to A_p for all m. Essentially, the z dependence in $\eta(z) = A_p/4\pi z^2$ is canceled by that of $M^2 = d^2/z^2$. Thus the detected photon density due to a broad area source at some plane z is independent of the distance z from the pinhole. This can be appreciated if we consider the photons collected by a point on the detector plane. As the source plane moves away, corresponding to increasing z, the collection efficiency goes down, but the source area seen by the point increases. These conflicting phenomena cancel and result in a z independence within the paraxial approximation. In general, with increasing z, the detected image becomes smaller as the photon efficiency decreases, maintaining the photon density.

The impulse response of equation (8.11) and the resultant intensity of equation (8.13) could also have been derived using the "alternative analysis using planar object" described in Chapter 4. As shown in Fig. 4.13, the response due to an impulsian pinhole can be formulated as in equation (4.37). The system impulse response is then derived by integrating over the entire aperture function. The desired system response is obtained by substituting the source distribution $S(x, y)$ for the x-ray source $s(x, y)$ and substituting the aperture function $a(x, y)$ for the transmission $t(x, y)$. The magnification constants m and M must be appropriately defined to achieve the desired result.

We have considered the intensity due to a specific plane. For the general case consisting of a volumetric source the resultant image is the linear superposition of the contribution due to all z planes in $S(x, y, z)$. The total integrated intensity is given by

$$I_d(x_d, y_d) = \frac{1}{4\pi d^2} \int \frac{1}{m(z)^2} \left[S\left(\frac{x_d}{M(z)}, \frac{y_d}{M(z)}, z\right) ** a\left(\frac{x_d}{m(z)}, \frac{y_d}{m(z)}\right) \right] dz. \quad (8.16)$$

This result is distinctly different from that of projection radiography, where the volumetric object resulted in the nonlinear relationships of (4.41) and (4.42). In those cases it was necessary to use various approximations to linearize the system and allow the use of convolutional forms.

Equation (8.16) has neglected the effects of the attenuation of the object. In general, each radiating source experiences the attenuation of the tissue before reaching the detector. Although the attenuation is relatively low at the high energies used, its effects can be considerable. One simplified approach is to assume that the tissue attenuation is a constant, equal to that of water. In that case, knowing the approximate outline of the region being studied, we can make a good estimate of the attenuation effect. For example, if the factor $e^{-\mu(z-z')}$ is included in equation (8.16), it will provide a significant attenuation correction. Here μ is the average attenuation coefficient of water or tissue at the energy used and z' is the distance of the border of the patient from the pinhole.

The size of the aperture represents a fundamental trade-off between resolution and efficiency. The total number of received photons is proportional to A_p. Assume that it is desired to improve the system resolution by reducing the linear size of a resolution element by a factor of 2. Since the number of elements in the detector has increased by four, four times as many detected photons are required to obtain the same statistics in each picture element. To improve the resolution of the imaging structure the width of the pinhole is halved, reducing the photon efficiency η by 4. Therefore, a 16:1 increase in the number of emitted photons are required to halve the linear size of a picture element. Thus the photon requirement varies as the fourth power of the linear resolution.

This fourth-power variation of the emitted photons with the linear dimensions of the aperture can be again seen by studying equation (8.14). This equation represents intensity or photons per unit area. If multiplied by the projected pinhole area, $m^2 A_p$, it will represent photons per element. The projected source $S(x/M, y/M)$ is therefore multiplied by A_p and convolved with the aperture function whose area is A_p. The required number of emitted photons, for a given number of photons per element, is therefore inversely proportional to A_p^2 or the fourth power of the linear extent of the aperture $a(x, y)$.

One significant difficulty with pinhole collimators is that the image magnification M varies inversely with depth. This can be significant in that lesions of unknown depth can appear with arbitrary magnification. This can be important in nuclear medicine, where there are few anatomical guidelines so that spatial distortions in an apparent lesion can be more serious. It is not apparent, therefore, whether a small image represents a small lesion or a large distant lesion. As a result of this deficiency, pinhole collimators are normally used for viewing organs at known depths such as the thyroid gland. They are rarely used for inspecting a volume, such as the head, for tumors that might occur anywhere in the space. A photograph of a commercial pinhole collimator together with a typical thyroid image are shown in Fig. 8.7.

FIG. 8.7 Pinhole collimator and a thyroid image made with the collimator.

PARALLEL HOLE COLLIMATOR

The parallel hole collimator [Anger, 1964] shown in Fig. 8.8 overcomes some of the problems of the pinhole collimator. In essence this structure is an attempt at collimating the emitted radiation so as to record an image having unity magnification at all depth planes. The region being studied is placed against the collimator so as to get all of the sources as close as possible. Unlike the pinhole the magnification does not vary with depth and is a constant at unity.

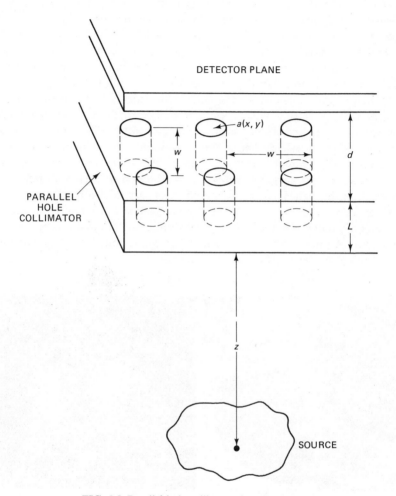

FIG. 8.8 Parallel-hole collimator imaging system.

As shown in Fig. 8.8, the parallel hole collimator is a block of high attenuation material, usually lead, with an array of spaced parallel holes each having an opening $a(x, y)$ and whose centers are separated by w in the x and y dimensions. The three-dimensional attenuation coefficient can be modeled as

$$\mu(x, y, z) = \bar{\mu}_0 \left[1 - a(x, y) ** \text{comb}\left(\frac{x}{w}\right) \text{comb}\left(\frac{y}{w}\right) \right] \text{rect}\left(\frac{z + L/2}{L}\right). \quad (8.17)$$

In general each hole of the collimator is responsive primarily to activity in its vicinity. This accounts for the unity magnification.

For a more detailed analysis we employ the general method suggested by Metz et al. [1980], where the collimator material is assumed to be perfectly opaque. We then explore the geometric projection of point sources at various positions through the holes. It is evident, observing Fig. 8.8, that this will lead to a space-variant formulation. Clearly, the response will be different for a point centered directly below a hole than one centered in a septal region between holes. Space-variant responses, although accurate, have questionable value in systems analysis since they do not result in convolutional forms, nor do they provide transfer functions in the frequency domain. For these reasons we use the concept of averaged responses as suggested in the Metz et al. paper. Since each source has uniform probability of being at each x, y position, we provide an inpulse response averaged over collimator position which is space invariant and has the desired properties.

The calculation of the impulse response is similar to that of the pinhole collimator except for the finite thickness of the collimator. The result is similar to that of equations (4.5) through (4.10), where μ_0 approaches infinity. This represents the rays reaching the detector which are not obscured by the collimator material. When the point source is at or near a region corresponding to the center of a hole, the impulse is simply the projection of the back aperture function nearest the detector, a magnified version of $a(x, y)$. As the point moves laterally, the rays begin to be obscurred by the front aperture function nearest the source, which has a larger magnification. The resultant response is the product of the two projections, as illustrated in Fig. 8.9.

Figure 8.9 shows the projection for a single on-axis hole. Photons can reach the detector only at the intersection of the projections of the back and front apertures. In studying the systems response we use the same notation as that of the pinhole collimator, where m is the magnification of the projection of the hole and M is the lateral magnification of the source position. It must be emphasized, however, that these magnifications are used solely in developing the system response. Clearly, the overall system magnification will be unity. Using the subscript 1 for the back aperture and 2 for the front aperture, we have

$$m_1 = \frac{z + L + d}{z + L} \tag{8.18}$$

$$m_2 = \frac{z + L + d}{z} \tag{8.19}$$

and

$$M_1 = -\frac{d}{z + L} \tag{8.20}$$

$$M_2 = -\frac{d + L}{z}. \tag{8.21}$$

For notational convenience, let the lateral aperture function in equation (8.17) be given by

$$b(x, y) = a(x, y) ** \text{comb}\left(\frac{x}{w}\right) \text{comb}\left(\frac{y}{w}\right). \tag{8.22}$$

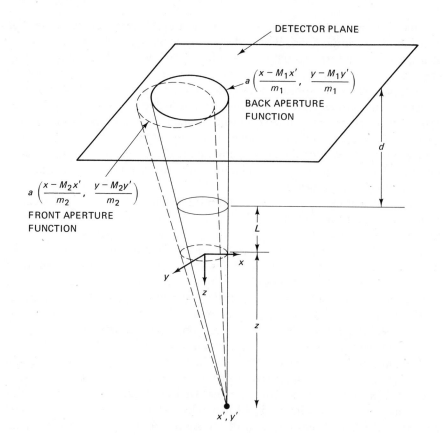

FIG. 8.9 Projection of a single-on-axis hole.

Using this notation the response at detector coordinates x_d, y_d to an impulse at x', y' at a depth z is given by

$$h(x_d, y_d, x', y') = \frac{\cos^3 \theta}{4\pi(z + L + d)^2} \, b\left(\frac{x_d - M_1 x'}{m_1}, \frac{y_d - M_1 y'}{m_1}\right)$$

$$\times \, b\left(\frac{x_d - M_2 x'}{m_2}, \frac{y_d - M_2 y'}{m_2}\right) \quad (8.23)$$

where, as before, θ is the obliquity angle of the rays from the normal. We again make the simplifying assumption that θ is sufficiently small to ignore $\cos^3 \theta$. This represents a stronger, more accurate, assumption than that of the pinhole collimator since only rays having small angles to the normal can penetrate a parallel hole collimator because of its thickness.

The impulse response of (8.23) is clearly space variant and cannot conveniently be used to find the image due to a general source distribution. Also, the affect of the collimator in the frequency domain cannot be studied. The space variance can be seen mathematically in that no change of variables can reduce the expression to represent the difference of input and output coordinates. This is confirmed physically in studying Fig. 8.9 in that different source positions produce different intersections of the projections of the front and back aperture functions.

As indicated, we will attempt to develop an average impulse response. In effect, we will move the collimator such that the origin moves over a period from $-w/2$ to $w/2$ in x and y and integrate the result. Our averaged impulse response is therefore

$$\bar{h}(x_d, y_d, x', y')$$

$$= \frac{\frac{1}{w^2} \int\limits_{-w/2}^{w/2}\!\!\int b\left(\frac{x_d - M_1 x' - m_1 x}{m_1}, \ldots\right) b\left(\frac{x_d - M_2 x' - m_2 x}{m_2}, \ldots\right) dx\,dy}{4\pi(z + L + d)^2} \quad (8.24)$$

where the y dependence in each expression for b is identical to that of x and has been omitted for brevity. Equation (8.24) can be structured into a convenient autocorrelation form using the substitutions

$$p = \frac{x_d}{m_1} - \frac{M_1}{m_1} x' - x, \qquad q = \frac{y_d}{m_1} - \frac{M_1}{m_1} y' - y \quad (8.25)$$

resulting in

$$\bar{h}(x_d, y_d, x', y') = \frac{1}{4\pi(z + L + d)^2 w^2} \int\limits_{-w/2}^{w/2}\!\!\int b(p, q)$$

$$\times b\left[p + \frac{L}{z + L + d}(x_d - x'), q + \frac{L}{z + L + d}(y_d - y')\right] dp\,dq \quad (8.26)$$

$$= \frac{1}{4\pi(z + L + d)^2 w^2} \Gamma_b\left[\frac{L}{z + L + d}(x_d - x', y_d - y')\right] \quad (8.27)$$

where $\Gamma_b(c, d)$ is the two-dimensional autocorrelation evaluated at c, d defined as

$$\Gamma_b(c, d) = \int\limits_{-w/2}^{w/2}\!\!\int b(p, q) b(p + c, q + d)\, dp\,dq. \quad (8.28)$$

Equation (8.27) clearly illustrates the averaged space-invariant impulse response. Since the impulse response is directly proportional to the difference between the source and detector coordinates the system has unity magnification.

The derivation was based on the lateral aperture function $b(x, y)$ in equation (8.22), consisting of a regular array of apertures $a(x, y)$. It is interesting to note that, under typical conditions, the intersection of the two aperture functions, as given in equations (8.24) and (8.26), includes just a single aperture. That is, a point source will produce an image only due to the single hole it is under, and none other. As we translate a point source from the axis, the area of overlap of the front- and back-projected aperture functions becomes less and less until it is zero. If the amount of translation at that point is less than $w/2$, half the distance between holes, then the point clearly cannot project through other holes. If it exceeds $w/2$, the impulse response will involve a number of holes.

If D represents the lateral extent of the hole $a(x, y)$, the condition for the impulse response to be based on a single aperture is given by

$$w > \frac{D(m_1 + m_2)}{M_1 - M_2} \tag{8.29}$$

corresponding to a depth range

$$z < \frac{L(w - D)}{2D}. \tag{8.30}$$

For most parallel hole collimators, typical depths will be within this range. Therefore, the average impulse response becomes

$$\bar{h}(x_d, y_d) = \frac{1}{4\pi(z + L + d)^2 w^2} \Gamma_a\left(\frac{L}{z + L + d} x_d, \frac{L}{z + L + d} y_d\right) \tag{8.31}$$

where Γ_a is the autocorrelation of the aperture $a(x, y)$. Using this impulse response, the intensity due to a general source distribution $S(x, y, z)$ is given by

$$I_d(x_d, y_d) = \int \frac{1}{4\pi(z + L + d)^2 w^2}$$
$$\times \left[\Gamma_a\left(\frac{L}{z + L + d} x_d, \frac{L}{z + L + d} y_d\right) ** S(x_d, y_d, z)\right] dz. \tag{8.32}$$

It must be recalled that this detected intensity is based on an averaged impulse response which would occur if the collimator were scanned during the exposure. It also represents the estimated impulse response of this space-variant system.

The efficiency is determined in the same way with the average efficiency $\bar{\eta}$ determined by the average projection area as given by

$$\bar{\eta} = \frac{1}{4\pi(z + L + d)^2 w^2} \iint \Gamma_a\left(\frac{L}{z + L + d} x_d, \frac{L}{z + L + d} y_d\right) dx_d dy_d. \tag{8.33}$$

Using the average impulse response we can develop a transfer function $\mathcal{C}(u, v)$ using the fact that the Fourier transform of an autocorrelation is the squared magnitude of the function's transform. The transfer function is given by

$$\mathcal{C}(u, v) = \mathcal{F}\{\bar{h}(x_d, y_d)\}$$
$$= \frac{1}{4\pi L^2 w^2}\left|A\left(\frac{z + L + d}{L} u, \frac{z + L + d}{L} v\right)\right|^2 \tag{8.34}$$

where $A(u, v)$ is the transform of the aperture function $a(x, y)$. This transfer function can be used to find the response due to any planar source as given by

$$I_d(u, v) = \mathcal{C}(u, v) S(u, v) \tag{8.35}$$

where $S(u, v)$ is the Fourier transform of a planar source distribution $S(x, y)$.

NOISE CONSIDERATIONS

The noise considerations in nuclear medicine imaging are dominated by the Poisson statistics of the relatively few detected photons. Scatter is an additional noise source because of the limited energy-selective capability of the detectors. The newer semiconductor detector systems have significantly greater energy resolution and thus provide greatly improved scatter rejection. This is often done, however, at some price in quantum efficiency.

Although the photon statistics are much poorer than that of radiography, the regions of interest can be delineated because of the greater contrast. For example, a brain lesion is difficult to detect radiographically since its attenuation coefficient differs only slightly from that of the surrounding tissue. As a result, the detection process requires large numbers of photons to reduce the standard deviation in the image. Conversely, in a nuclear medicine procedure the lesion takes up much more of the isotope than the surrounding tissue so that many fewer photons are required to distinguish the lesion.

Assuming that we have a source emitting a background level of n counts per unit area, the noise or standard deviation of the measurement is given by

$$\sigma = \sqrt{\frac{\eta}{M^2} nA} \tag{8.36}$$

where A is the area of a picture element in the image and $\eta n / M^2$ is the photon density at the image. The signal can be determined as in radiography, where C is the fractional difference between the photon density at the area of interest and the background resulting in a signal-to-noise ratio given by

$$\text{SNR} = C\sqrt{\frac{\eta}{M^2} nA}. \tag{8.37}$$

This assumes that the amplitude of the signal in the region of interest is unaffected by the blurring of the impulse response. A more accurate representation involves defining the image signal as the difference in the number of photons per element at the center of the lesion and in the background. We therefore convolve the image with the impulse response, as in equations (8.14) and (8.31), and evaluate the convolution at the center of the lesion. This is equivalent to taking the integral of the product of the image distribution and the impulse response centered at the center of the lesion. The difference between this value in photons per picture element and the background determines the signal.

The SNR is an interesting function of the aperture size in a pinhole imaging system. For example, assume that we are imaging a "cold" lesion such as occurs in the liver. This can be represented by a uniformly emitting area with a small nonemitting region representing the lesion. For a relatively small pinhole, the cold lesion is well resolved, so that the signal is equal to the background level since the value at the center of the lesion is zero. However, the low efficiency of

163

the small pinhole results in a low photon count and low SNR. As we increase the pinhole size, the photon count and SNR increases. However, at a certain aperture size, the lesion is no longer well resolved, so that the center of the lesion, due to convolution with the aperture function, is no longer zero. The reduced signal can then decrease the SNR. Clearly, there is an optimum size of pinhole aperture for each lesion size.

The system can be analyzed assuming a planar source distribution consisting of a background photon density $b(x, y)$ and a "lesion" density $l(x, y)$ centered at x_0, y_0 as given by

$$S(x, y) = b(x, y) + l(x - x_0, y - y_0). \tag{8.38}$$

The signal is defined as the magnitude of the difference in photons per pixel, between the center of the lesion and the background. The impulse response $h(x_d, y_d)$ is used to determine the response at each region as given by

$$\text{signal} = A \int\int h(x_d - Mx_0, y_d - My_0) l\left(\frac{x_d}{M} - x_0, \frac{y_d}{M} - y_0\right) dx_d dy_d \tag{8.39}$$

where A is the pixel area and M the magnification of the plane. The noise, as previously indicated, is the standard deviation of the background signal:

$$\text{noise} = \sqrt{A \int\int h(x_d, y_d) b\left(\frac{x_d}{M}, \frac{y_d}{M}\right) dx_d dy_d} \tag{8.40}$$

which, for $b(x, y)$ equal to a constant n photons per unit area, becomes $\sqrt{\eta n A / M^2}$.

CODED APERTURE SYSTEMS

In recent years there has been considerable research effort at imaging structures with increased capture efficiency as compared to the 10^{-4} of pinhole and parallel hole collimators. One approach is the use of an imaging structure consisting of an array of pinholes known as a coded aperture plate [Barrett, 1972; Barrett et al., 1972]. This structure, shown in Fig. 8.10, is used in place of the pinhole in a

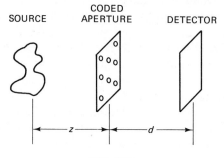

FIG. 8.10

nuclear medicine camera. The resultant coded image is thus the convolution of the source with that of the aperture plate. This image must then be decoded, by a suitably magnified version of the aperture plate, to reconstruct the object at any plane. Since the convolution is a function of the depth plane, three-dimensional information is derived. In addition, the average transmission of the aperture plate can be made many times greater than that of the pinhole. Using the same notation as the pinhole analysis, the detected intensity due to a source at plane z is given by

$$I_d = \frac{1}{4\pi d^2 m^2} \, S\left(\frac{x_d}{M}, \frac{y_d}{M}\right) ** a\left(\frac{x_d}{m}, \frac{y_d}{m}\right) \tag{8.41}$$

where $S(x, y)$ is the source distribution at plane z, $a(x, y)$ is the transmission of the aperture plate, and

$$M = -\frac{d}{z} \quad \text{and} \quad m = \frac{z + d}{z} = 1 - M. \tag{8.42}$$

In decoding the detected image, we cross correlate with the appropriately magnified version of the aperture plate to provide the reconstructed image as given by

$$I_r = I_d \star\star a\left(\frac{x_d}{m}, \frac{y_d}{m}\right) \tag{8.43}$$

$$= \frac{1}{4\pi d^2 m^2} S\left(\frac{x_d}{M}, \frac{y_d}{M}\right) ** \left[a\left(\frac{x_d}{m}, \frac{y_d}{m}\right) \star\star a\left(\frac{x_d}{m}, \frac{y_d}{m}\right)\right]. \tag{8.44}$$

The point response of the reconstructed image $h(x, y)$ is given by the autocorrelation of the aperture function as

$$h(x_d, y_d) = a\left(\frac{x_d}{m}, \frac{y_d}{m}\right) \star\star a\left(\frac{x_d}{m}, \frac{y_d}{m}\right). \tag{8.45}$$

Thus an aperture plate should be chosen which has a high-amplitude, narrow autocorrelation peak for good resolution and large open area for good efficiency. Examples are random pinhole arrays and Fresnel zone plates. The latter can be reconstructed optically since propagation through space provides the desired correlation function. A different magnification m is used for each depth plane so as to reconstruct each depth region separately.

This imaging structure exhibits good performance for small sources. For larger sources, however, comparable to the size of the coded aperture, the noise performance is considerably degraded. The basic problem arises from the non-negative nature of the aperture function $a(x, y)$. Its autocorrelation, for an aperture containing a large number of holes, will consist of a narrow central peak having a normalized value of n, the number of holes, and a large background pedestal having a value of about unity, corresponding to the overlap of single holes in the autocorrelation. The convolution with this function essentially produces an image that is amplified by n, plus an integral of the image as a result of the large background pedestal of the function.

It is instructive to compare the signal-to-noise ratio of a single pinhole system to that of a coded aperture with n pinholes each having the same size. The signal-to-noise ratio of a uniform region using the single pinhole is given by

$$\text{SNR}_{\text{pinhole}} = \frac{N}{\sqrt{N}} = \sqrt{N} \tag{8.46}$$

where N is the number of detected photons per picture element. In the multiple aperture plate the signal is given by nN because of the autocorrelation function. The variance is the sum of the noise due to the narrow peak in the autocorrelation and the integrated sum of the sources due to the large pedastal. Since these are independent, the variance is given by

$$(\text{variance})_{\text{multiple aperture}} = nN + mN \tag{8.47}$$

where m is the number of equal intensity sources. The resultant signal-to-noise ratio is given by

$$\text{SNR}_{\text{multiple aperture}} = \frac{nN}{\sqrt{nN + mN}} = \frac{n\sqrt{N}}{\sqrt{n + m}}. \tag{8.48}$$

For small sources, where $n \gg m$, this becomes

$$\text{SNR}_{\text{multiple aperture}} \simeq \sqrt{n}\sqrt{N} = \sqrt{n}\,\text{SNR}_{\text{pinhole}} \tag{8.49}$$

where the improvement is obvious. For large sources $m \gg n$ the signal-to-noise ratio significantly deteriorates. For highly extended sources the resultant signal-to-noise ratio is poorer than that of the single pinhole so that the coded aperture system results in poorer noise performance. However, it continues to provide depth information which the single pinhole does not. Improved versions of these imaging systems are presently under study.

TOMOGRAPHIC SOURCE IMAGING

To avoid dealing with a projection of a volumetric object, tomographic systems are used to provide three-dimensional information. These have direct analogies with x-ray tomographic systems, so that we can rely heavily on the results of Chapter 7. For example, a simple motion tomography system can be created from the pinhole imaging system of Fig. 8.6 by moving the pinhole in a line pattern described by $f(x, y)$ with the detector moved in a magnified pattern $f(x/m, y/m)$. This will result in the image at plane z remaining in focus and the others being blurred by varying amounts.

As with radiography, motion tomography provides limited improvement since the intermediate planes are still present. Computerized tomography provides isolated sections of the three-dimensional volume completely free of intermediate structures [Budinger and Gullberg, 1974]. Here the data are acquired with a basic nuclear medicine camera, usually using a parallel hole collimator, moved around the patient to collect an array of projection images at many angles. These data are processed exactly as described in Chapter 7, where each measurement represents the line integrals of the sources at a particular angle. The reconstruction usually uses the convolution–back projection algorithm described in detail in Chapter 7. Because of the sensitivity of these reconstruction systems, it is particularly important to correct for attenuation so as to obtain the true line integrals of the source distribution. This can be accomplished by assuming a fixed value of μ as previously indicated. For greater accuracy, however, we can use an external source at the same energy as the isotope being imaged to provide a reconstruction of $\mu(x, y)$ using transmission computerized tomography. These values are then used to correct for the measured projections of the source distribution.

In nuclear medicine, using a parallel hole collimator, at each projection angle we simultaneously measure the projections of an array of planar sections. The complete reconstruction is therefore an array of continuous sections of the volume. A typical series of reconstructions of the brain is shown in Fig. 8.11.

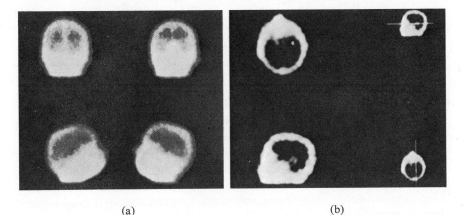

(a)   (b)

FIG. 8.11 Conventional and cross-sectional reconstruction images of the brain. (Courtesy of the General Electric Medical Systems Division.)

In this study conventional projection images of the activity of the brain are compared to cross-sectional reconstruction. The patient had a lesion in the left temporal lobe which was not visible on the projection images. To provide the cross-sectional reconstruction, 64 views were acquired at a rate of 30 seconds per view. These were reconstructed and examined. The lesion appears on the transverse and sagittal sections shown. It is clear that in this case the projection images, representing the superimposed activity of all planes, failed to demonstrate the disease.

POSITRON IMAGING

Positron emitters generate a unique configuration of gamma rays. Each emitted positron almost immediately interacts with an electron to produce an annihilation event which generates two gamma rays each having energies of 510 kev and at almost exactly opposite directions [Meredith and Massey, 1977; Johns and Cunningham, 1974]. This phenomenon gives rise to a camera system shown in Fig. 8.12. Here a pair of two-dimensional position-indicating detectors are used

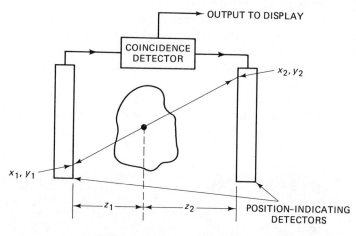

FIG. 8.12 Positron imaging system.

on either side of the subject. These can be either discrete arrays of detectors or Anger cameras as previously described. The energy-selective mechanisms of the detectors are set for 510 kev. When an annihilation event occurs, two photons travel to the individual detectors. A coincidence detector records an output event only in response to gamma rays being received at both detectors. This helps to eliminate various undesired events, such as Compton-scattered photons. Following a coincidence, the source position can be estimated for a known depth plane as given by

$$\hat{x} = \frac{x_1 z_2}{z_1 + z_2} + \frac{x_2 z_1}{z_1 + z_2}, \qquad \hat{y} = \frac{y_1 z_2}{z_1 + z_2} + \frac{y_2 z_1}{z_1 + z_2} \qquad (8.50)$$

where x_1, y_1 and x_2, y_2 are the coordinates of the first and second detectors and z_1 and z_2 are the respective distances from the detectors to the source point. In general, a given plane of interest is studied so that the reconstruction is accurate in that plane and blurred for other planes as with motion tomography. The significant feature is that no collimators are required. The only limitation to the photon collection angle is the size of the detecting planes themselves.

One method of avoiding the blurring from other planes is to measure the difference of arrival time of the pair of photons and use that information to determine the source plane. Unfortunately, considering the velocity of light, it would require 30-picosend accuracy to provide 1.0-cm-depth resolution. That is well beyond our present electronic capability, both in the detection and processing systems. This method does, however, have interesting potential for the future.

A recent exciting approach to the reconstruction, which avoids the overlap of planes, is the line integral technique discussed in Chapter 7 in the section on computerized tomography. If we sum up all the coincidence events reaching each pair of detector locations x_1, y_1 and x_2, y_2, we will have calculated the line integral of all the sources along that line. Having the line integral measurement over all angles and positions, we can then reconstruct the complete source distribution.

A system for accomplishing that specific task is the positron ring shown in Fig. 8.13 [Ter-Pogossian et al., 1975]. This system is used to reconstruct the

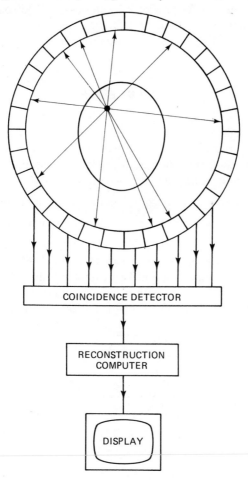

FIG. 8.13 Positron ring detector.

sources in a plane. The sum of coincident events in each detector pair around the ring represents the desired line integral. Since the line integral is available for a complete range of angles and positions, the reconstruction techniques used can be the same as those discussed in Chapter 7. The attenuation should again be considered, although, at these higher energies of 510 kev, it is somewhat less of a problem.

Positron imaging has two inherent limitations to its ultimate resolution. These are the range of the positron between its point of creation at the isotype and its point of annihilation, and the departure from 180° of the two photons because of the momentum of the positron. It should be emphasized, however, that in present systems the position detection accuracy, photon statistics, and sampling are the major limitations which provide a lateral resolution in the order of 1.0 cm. One practical difficulty in the use of these cameras is that many of the desirable isotopes must be produced by an on-site cyclotron. However, some of the resultant images represent outstanding delineations of perfusion and metabolism which are unavailable using other modalities.

PROBLEMS

8.1 In a pinhole source imaging system, the aperture $a(x, y) = \text{rect}(x/X)$ rect (y/Y). During the exposure the aperture center is translated from $x = -D_1/2$ to $x = D_1/2$ while the detector center is translated from $x_d = -D_2/2$ to $x_d = D_2/2$ in the same direction, where $D_2 > D_1$.

(a) At what plane is the image not degrated by the motions?

(b) Find the resultant recorded point response function $h(x_d, y_d)$ as a function of depth z.

8.2 A source consists of an infinite sheet of isotropically radiating material emitting n photons/unit area and has a nonemitting hole of radius a (Fig. P8.2). The source is imaged with a circular pinhole of radius b as shown using a pixel area A. The signal is defined as the difference in the number of

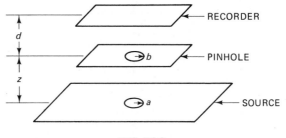

FIG. P8.2

photons per pixel between the background level of the image and the level at the center of the hole.

(a) Find the signal-to-noise ratio versus b. [*Hint:* Use analytic expressions for two ranges of b.]

(b) Find the pinhole size that maximizes the signal-to-noise ratio.

8.3 A source consists of an infinite sheet of isotropically radiating material emitting n photons per unit area and has a nonemitting rectangular hole $D \times 2D$. The source is imaged with a $B \times B$ square pinhole a distance z from the source and d from the detector. The signal is defined as the difference in photons per pixel between the background and the center of the hole image. The area of a pixel is A.

(a) Calculate SNR versus B.

(b) Find the size of B for optimum SNR.

8.4 An infinite planar source having a photon emission density $n_0(1 + p \cos 2\pi f_0 x)$ is imaged using a pinhole system with a source to pinhole distance z and a pinhole to detector distance d. The signal is defined as the peak of the number of photons per pixel of the detected sinusoid. Assume a pixel area of A.

(a) Find an expression for the signal-to-noise ratio for a general aperture $a(x, y)$.

(b) For an aperture having a Gaussian transmission, $e^{-(r/b)^2}$, find the value of b that maximizes the signal-to-noise ratio.

8.5 The same source as in Problem 8.4 is imaged using a parallel hole collimator having a thickness L with circular holes of diameter D. Calculate the signal-to-noise ratio assuming that the impulse response is limited to the projection of a single hole.

8.6 As shown in Fig. P8.6, a pinhole imaging system of radius R is used to image a volumetric source distribution of thickness L, having a cylindrical

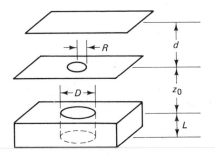

FIG. P8.6

hole of diameter D, where $D/2 > R$, and emitting n photons per unit volume. The signal is defined as the difference in photons per pixel of the background level and the center of the hole image. Assuming a pixel area of A, calculate the SNR versus z_0, the depth of the emitting source.

8.7 A parallel hole collimator has $D \times D$ square holes whose centers are separated by w in both dimensions, a thickness L with the detector plane a distance d from the top of the collimator. A planar source a distance z_0 from the bottom of the collimator has a distribution of $n_0[1 - \text{rect}(x/B) \text{rect}(y/B)]$ photons per unit area. Assume the hole separation is such that the impulse response at z_0 involves a single hole. Using the average impulse response and a pixel area of A, calculate the SNR.

9

Basic Ultrasonic Imaging

In this chapter the basic concepts of ultrasonic imaging [Wells, 1969; Woodcock, 1979] are introduced using a simplified model and some reasonable approximations. Although these simplifications and approximations lead to some inaccuracies, they do form the basis of most current medical ultrasonic imaging instruments.

This chapter is limited to the reflection imaging modality, where, as in radar, ultrasonic pulses are propagated through the body, causing reflected waves to occur at various discontinuities throughout the path of the propagated beam. Reflection or echo imaging is thus far the only one that has achieved commercial use. Other ultrasonic imaging modalities, which have thus far achieved only experimental use, are considered in Chapter 11.

This imaging modality is made possible by the relatively slow velocity of propagation of about 1500 meters/sec. This represents about a 333-μsec round trip time through 25 cm. In this time scale it is relatively simple for modern electronic circuitry to distinguish reflections at different depths with good resolution. This is in sharp distinction to the x-ray region, where the energy travels at the speed of light, 3×10^8 meters/sec. At these speeds it would require picosecond accuracies to distinguish various depths in the body. Current electronic techniques have not yet reached this capability. As a result, x-ray imaging, as has been described, is limited to the transmission modality.

Another fundamental characteristic of reflection ultrasonic imaging is the direct acquisition of three-dimensional information. X-ray imaging systems basically acquire projection information or line integrals of the attenuation coefficient. In these systems three-dimensional information can be acquired only indirectly through computer reconstructions of many projections, as described in Chapter 7. In ultrasound, however, the received signal directly indicates the reflectivity of the object in three dimensions. The propagating beam pattern defines the lateral coordinates, and the round-trip time of the reflected pulse defines the depth coordinate. Thus each received pulse directly represents the reflectivity at a point in object space.

In this chapter we use many assumptions and approximations on the characteristics of the volume of the body being studied and on the nature of the propagating wave. In each case we indicate the degree to which these assumptions are valid. In the case of the propagating wave we include three analyses, each having a different degree of accuracy. This approach will be continued into the next chapter in the study of arrays.

BASIC REFLECTION IMAGING

A basic reflection imaging arrangement is illustrated in Fig. 9.1. With the switch thrown in the transmit position, the pulse waveform $p(t)$ excites the transducer, resulting in the propagated wavefronts shown in the solid lines. Immediately following the pulse transmission the switch is thrown into the receive position using the same transducer. When the wavefront hits a discontinuity, as shown, a scattered wave is produced having the directions indicated by the dashed lines. This scattered wave is received by the same transducer and the resultant signal is processed and displayed. The processing usually consists of bandpass filtering, gain control, and envelope detection.

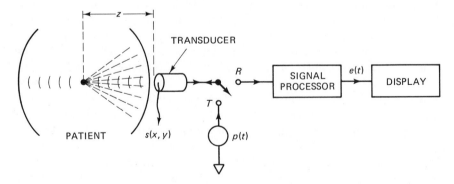

FIG. 9.1 Basic reflection imaging system.

As an initial approximation we assume that the diameter or extent of the face of the transducer is very large as compared to the wavelength of the propagating wave. Under these circumstances the propagating wave approaches a geometric extension of the transducer face $s(x, y)$. Diffraction spreading can be ignored under these circumstances. We also assume that the wave propagates with a velocity c which is uniform throughout the body and is attenuated with an attenuation coefficient α which is also uniform. If we model the body as an array of isotropic scatterers [Nicholas, 1977] having a reflectivity $R(x, y, z)$, the resultant processed signal $e(t)$ is given by

$$e(t) = K \left| \int \int \int \frac{e^{-2\alpha z}}{z} R(x, y, z) s(x, y) \tilde{p} \left(t - \frac{2z}{c} \right) dx dy dz \right| \qquad (9.1)$$

where K is a normalizing constant, $e^{-2\alpha z}$ is the attenuation in the tissue through the round-trip distance of $2z$, $s(x, y)$ is the lateral distribution of the propagating wave, and $\tilde{p}(t - 2z/c)$ is the received pulse delayed by the round-trip time $2z/c$ and modified by the various linear processes. Thus $\tilde{p}(t)$, the received pulse, is the convolution of the transmitter pulse $p(t)$ with the impulse responses of the transducer and the associated linear filters in the processor. It also includes a derivative operator which is basic to the propagation phenomena, since it is the change in the ultrasonic parameters, such as pressure, which gives rise to the propagating wave. The absolute value used represents the envelope detection, which is phase insensitive. The $1/z$ factor is the loss in amplitude of the reflected wave due to diffraction spreading from each scatterer as illustrated in Fig. 9.1.

In general, equation (9.1) should have the transducer characteristic $s(x, y)$ squared since it represents both the transmitted and received pattern. Thus $s(x, y)$ is both the lateral insonification function and the lateral receiver sensitivity function. However, in this simplified example, where we ignore diffraction, we are also assuming that $s(x, y)$ is constant over the transducer face and zero elsewhere. This is a reasonable approximation for most transducers which are operating in the piston mode.

$R(x, y, z)$ has been assumed to be a scalar in that the reflectivity is independent of the angle of approach of the ultrasonic beam [Nicholas, 1977]. This is accurate for structures that are small compared to a wavelength and thus approach isotropic scattering. It is also accurate for relatively large areas whose rms roughness is large as compared to a wavelength and thus become diffusely scattering in an almost isotropic fashion. For relatively smooth large surfaces, however, which give rise to specular reflections, the model is inaccurate since the reflection depends strongly on the angle of approach of the beam and thus becomes a complex vector problem. This will be considered later in the chapter.

Another assumption on the reflectivity $R(x, y, z)$ is that it is a *weakly* reflecting medium. This essentially refers to reflections which are sufficiently small that second-order reflections can be ignored. Otherwise, equation (9.1) would become significantly more complex so as to include the multiple bounces from various scatterers. This assumption is quite good in practice since the reflectance of most biological structures is quite small. Multiple reflections can occur at interfaces between tissue and bone, or tissue and air where strong reflections occur. These areas, however, have a variety of ultrasonic imaging problems and are generally considered unsuitable. Various properties of tissue, including reflectivity, velocity, and attenuation, will be presented toward the end of this chapter.

ATTENUATION CORRECTION

To simplify equation (9.1) we make use of the fact that the attenuation functions $e^{-2\alpha z}$ and $1/z$ vary relatively slowly with z. The received pulse $\tilde{p}(2z/c)$, however, is a relatively narrow pulse and occupies only a narrow depth range as it propagates. This is essential for good depth resolution. As a result the function $\tilde{p}(t - 2z/c)$ in equation (9.1) acts as a delta function as far as $e^{-2\alpha z}/z$ is concerned, resulting in the approximate output signal

$$e(t) \simeq K \left| \frac{e^{-\alpha ct}}{ct/2} \int \int \int R(x, y, z)s(x, y)\tilde{p}\left(t - \frac{2z}{c}\right)dxdydz \right| \qquad (9.2)$$

In essentially all ultrasonic imaging systems the signal processor includes a system of time-varying gain to compensate for the attenuation and diffraction spreading. Thus a compensated output signal $e_c(t)$ is produced which is the original output $e(t)$ multiplied by the time-varying gain $g(t) = cte^{\alpha ct}$ as given by

$$e_c(t) = g(t)e(t) = cte^{\alpha ct}e(t). \qquad (9.3)$$

Using the time-varying gain the compensated output signal becomes

$$e_c(t) = K \left| \int \int \int R(x, y, z)s(x, y)\tilde{p}\left(t - \frac{2z}{c}\right)dxdydz \right|. \qquad (9.4)$$

It is convenient and instructive to structure this resultant signal in convolutional form as given by

$$e_c(t) = K \left| R\left(x, y, \frac{ct}{2}\right) *** s(-x, -y)\tilde{p}(t) \right| \qquad \text{evaluated at } x = 0, y = 0 \quad (9.5)$$

where the triple asterisk represents a three-dimensional convolution.

THE A SCAN

The signal $e_c(t)$ should ideally represent the reflectivity, as a function of time, along the $x = 0$, $y = 0$ axis of the body with time representing the various depths. In practice, this signal is used to deflect the beam of a synchronously driven display device with the beam deflection representing reflectivity versus depth [Wells, 1969]. What is actually displayed is an estimate of the reflectivity $\hat{R}(0, 0, z)$ which is obtained by scanning the display at the velocity $c/2$ as given by

$$\hat{R}(0, 0, z) = \int e_c(t)\delta\left(t - \frac{2z}{c}\right) dt$$

$$\hat{R}(0, 0, z) = K\left| R(x, y, z) *** \tilde{p}\left(\frac{2z}{c}\right) s(x, y)\right| \quad \text{evaluated at } x = 0, y = 0.$$

(9.6)

To be precise (9.6) should use $s(-x, -y)$ in the convolution relationship. However, rather than carry around this awkward notation, we can assume that $s(x, y)$ refers to an inverted source pattern. This represents no problem since, in both our examples and in commercial practice, symmetrical transducers are used where $s(x, y) = s(-x, -y)$. The estimate of the reflectivity along any other line, $\hat{R}(x_0, y_0, z)$, is found by simply moving the transducer to the point x_0, y_0 with the convolution evaluated as $x = x_0$ and $y = y_0$. Here we see the fundamental resolution limits of ultrasonic imaging where the lateral resolution is limited by the beam pattern $s(x, y)$ and the axial or depth resolution is limited by the received pulse waveform $\tilde{p}(2z/c)$. A rectangular pulse, $\tilde{p}(t) = \text{rect}(t/\tau)$, results in a depth response of $\text{rect}(2z/c\tau)$. The volumetric resolution element is the product of these lateral and depth functions.

This display of reflectivity as a function of depth, by deflecting the beam of a cathode ray tube, is known as an *A scan*. A typical A scan of the eye is shown in Fig. 9.2. These are widely used to study various other disease processes, including head injuries. In cases of head injury these scans are often used to find the position of the brain midline. A shift of this midline position can indicate bleeding and the need for urgent surgery.

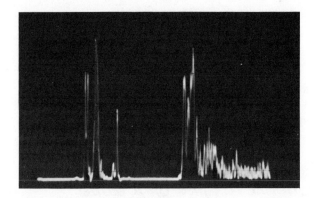

FIG. 9.2 An A scan of an eye which is normal except for a mild cataract in the lens. The first echo on the left is due to the cornea, with the next two representing the front and back of the lens. The small echo within the lens is the cataract. The next echo is the back of the eye, followed by an array of echoes due to retrobullar fat.

THE M MODE

We are often interested in the study of time-varying regions, such as the heart, where the anatomy changes relatively rapidly [Woodcock, 1979]. This can simply be modeled as a time-varying reflectivity function $R(x, y, z, t)$. A sequence of repetitive A scans are recorded separated by time T, where

$$T > \frac{2z_{max}}{c} \qquad (9.7)$$

where $2z_{max}/c$ is the round-trip propagation time to the maximum depth z_{max}. The compensated output signal $e_c(t)$ is then given by

$$e_c(t) = K \sum_{n=0}^{N} \left| \int \int \int R(x, y, z, t)s(x, y)\tilde{p}\left(t - nT - \frac{2z}{c}\right) dx dy dz \right| \qquad (9.8)$$

where a total of $N + 1$ lines are recorded in a time interval $(N + 1)T$.

As with the attenuation correction, we can assume with reasonable accuracy that the anatomy is stationary over each round-trip time $2z_{max}/c$. The round-trip time of about 300 μsec is very small compared to any anatomical motion. Therefore, equation (9.8) can be rewritten as

$$e_c(t) = K \sum_{n=0}^{N} \left| \int \int \int R(x, y, z, nT)s(x, y)\tilde{p}\left(t - nT - \frac{2z}{c}\right) dx dy dz \right| \cdot \qquad (9.9)$$

where t is approximated by nT.

This sequence of scans are displayed as intensity modulations of a display rather than deflection as with the A scan. The scans are arranged in a raster configuration so that, for example, the y axis indicates depth z and the x axis the time nT. This display represents an estimate of the time-varying reflectivity and consists of a sequence of A scans as given by

$$\hat{R}(0, 0, z, t) = K \sum_{n=0}^{N} \left| R(x, y, z, nT) *** \tilde{p}\left(\frac{2z}{c}\right) s(x, y) \right|$$

$$\text{evaluated at } x = 0, y = 0 \qquad (9.10)$$

where the convolution is with respect to the spatial coordinates x, y, and z.

A typical *M-mode scan* of the mitral valve is shown in Fig. 9.3, where the y axis represents z or depth and the x axis the quantized time nT. The total time corresponds to about four heartbeats.

CROSS-SECTIONAL IMAGING OR B MODE

The most popular presentation is the *B scan* or *B mode*, which is the reflectivity of a two-dimensional slice through a portion of the anatomy [Woodcock, 1979].

These data are usually acquired by linearly scanning the transducer of Fig. 9.1 at a uniform velocity. For example, assume that we wish to form a cross-

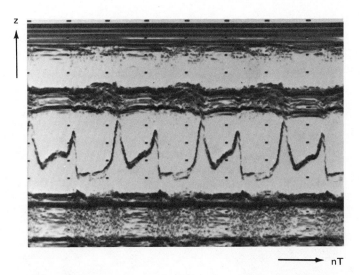

z

nT

FIG. 9.3 M-mode scan of the heart in the region of the mitral valve. The valve leaflet is shown undergoing significant motion during each heartbeat. The dark bands immediately above the initial valve are the reflections from the septum, separating the left and right chambers of the heart. This region undergoes negligible motion. (Courtesy of the General Electric Medical Systems Division.)

sectional image of the $y = y_0$ plane. The transducer is translated in the x direction at a uniform velocity v along $y = y_0$. As with the M mode, a sequence of scans of time T are produced as the transducer is translated. The compensated output is given by

$$e_c(t) = K \sum_{n=0}^{N} \left| \int \int \int R(x, y, z)s(x - vt, y - y_0)\tilde{p}\left(t - nT - \frac{2z}{c}\right) dx\,dy\,dz \right| \cdot$$
(9.11)

The reflectivity $R(x, y, z)$ is assumed to be constant during the generation of the image since little or no motion takes place in the body.

The transducer is assumed to be essentially stationary during each round-trip time T, providing an output signal

$$e_c(t) \simeq K \sum_{n=0}^{N} \left| \int \int \int R(x, y, z)s(x - vnT, y - y_0)\tilde{p}\left(t - nT - \frac{2z}{c}\right) dx\,dy\,dz \right| \cdot$$
(9.12)

As with the M mode, the output signal from each scan line is used to intensity-modulate a line in a raster display. This display provides an estimate of the reflectivity in the y_0 plane as given by

$$\hat{R}(x, y_0, z) = K \sum_{n=0}^{N} \left| R(x, y, z) *** s(x, y)\tilde{p}\left(\frac{2z}{c}\right) \right|$$

evaluated at $x = vnT$ and $y = y_0$. (9.13)

The B scan represents, by far, the most widely used modality in ultrasonic imaging. It provides a direct representation of the cross-sectional anatomy which can be readily evaluated. A typical B-scan system with a manually translated arm is shown in Fig. 9.4.

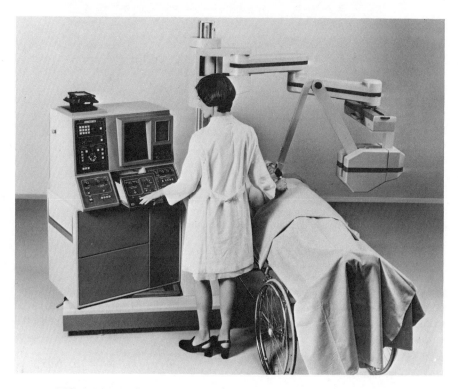

FIG. 9.4 Manually scanned *B*-mode system. (Courtesy of Siemens Gammasonics, Inc., Vetrasound Division.)

An abdominal B scan is shown in Fig. 9.5. This is an image of the liver where the upper border is the skin line along which the transducer was scanned. The hepatic vein is clearly shown. The curved lower boundary is the diaphragm.

This analysis was made with certain assumptions about the ultrasonic model of the region of the body being studied and of the nature of the propagation phenomenon. The region was assumed to have a constant velocity of propagation c, constant attenuation α, and composed of an array of weakly reflecting isotropic scatterers. The propagation phenomenon of the transmitted wave was assumed to be governed by geometric optics with diffraction neglected. We now proceed to examine these more closely and to refine them where appropriate.

We first consider the important problem of diffraction spreading and then consider our ultrasonic model of tissue.

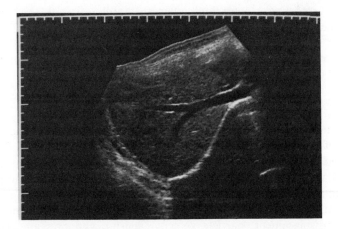

FIG. 9.5 B scan of the liver. (Courtesy of the General Electric Systems Division.)

DIFFRACTION FORMULATION

Diffraction spreading [Goodman, 1968; Norton, 1976] due to the relatively large ultrasonic wavelength represents the major factor determining the resolution limits in ultrasonic imaging. In studying the diffraction problem we consider the propagation, ignoring attenuation, between a point on the transducer in the x_0, y_0 plane and a point at depth z in the x_z, y_z plane as shown in Fig. 9.6. The basic propagation delay is modeled by $\delta(t - r_{0z}/c)$, indicating a delay of

$$\frac{r_{0z}}{c} \quad \text{where } r_{0z} = \sqrt{z^2 + (x_0 - x_z)^2 + (y_0 - y_z)^2}.$$

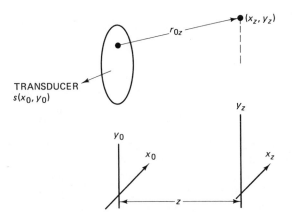

FIG. 9.6 Basic propagation model.

A number of linear operations affect the resultant waveform, including the derivative operation which is basic to the propagation phenomenon and the transducer characteristics. For convenience these are combined into a single impulse response $a(t)$. Thus the overall impulse response representing the signal received is given by

$$h(x_0, y_0; x_z, y_z; t) = \left[\delta\left(t - \frac{r_{0z}}{c}\right) * a(t) \right]\frac{z}{r_{0z}^2} \qquad (9.14)$$

where the z/r_{0z}^2 term represents the product of an obliquity factor z/r_{0z}, the cosine of the angle of incidence, and $1/r_{0z}$, the normal falloff with distance of an isotropic radiator.

We will use the impulse response h to find the field amplitude at any plane z due to a signal at the transducer plane where $z = 0$. This field amplitude can represent any of the following incremental parameters of the medium, including particle displacement, particle velocity, density, and pressure. These are all small-signal quantities representing departures from larger equilibrium values. The derivative operation previously referred to as a component of the linear filtering $a(t)$ indicates that a change in a parameter is required for propagation to occur. For example, the pressure measured at some point is proportional to the rate of change of pressure at a source point.

Consider a transducer at the origin driven by a sinusoidal burst providing a field amplitude given by

$$u(x_0, y_0, t) = s(x_0, y_0)p(t) \exp(-i\omega_0 t) \qquad (9.15)$$

where $s(x_0, y_0)$ is the spatial amplitude distribution at the transducer, $p(t)$ is the pulse envelope, and $\exp(-i\omega_0 t)$ is the sinusoidal carrier at a frequency ω_0. The field amplitude at any point in plane z due to a radiating point on the transducer is found by convolution with the impulse response of equation (9.14). The total field amplitude at plane z, $u(x_z, y_z, t)$, is found by convolving with equation (9.14) and integrating over the entire transducer plane as given by

$$u(x_z, y_z, t) = \int\int u(x_0, y_0, t) * h(t)dx_0 dy_0 \qquad (9.16)$$

where the convolution is with respect to time. Substituting for $u(x_0, y_0, t)$ and performing the convolution gives

$$u(x_z, y_z, t) = \int\int s(x_0, y_0)p\left(t - \frac{r_{0z}}{c}\right) \exp(ikr_{0z})\left(\frac{z}{r_{0z}^2}\right) dx_0 dy_0$$
$$\times \exp(-i\omega_0 t) * a(t) \qquad (9.17)$$

where the wavenumber $k = \omega_0/c = 2\pi/\lambda$. Equation (9.17) thus represents the insonification pattern at plane z. The integral operation provides the lateral extent of the pattern or its lateral resolution. The temporal functions define the depth resolution, as will be shown.

The insonification of each point at plane z is $u(x_z, y_z, t)$. We now study the response from the reflected wave back to the transducer. Assuming a unity reflecting point at x_z, y_z, the received signal $e_h(t)$ is given by

$$e_h(x_z, y_z, t) = \int\int u(x_z, y_z, t) * \left[\delta\left(t - \frac{r_{0z}}{c}\right) * b(t) \right]\frac{z}{r_{0z}^2} s(x_0, y_0) dx_0 dy_0 \quad (9.18)$$

where $\delta(t - r_{0z}/c)$ is the impulse response from the reflecting point to each point in the transducer, $b(t)$ represents the linear operations between the reflecting point and the received electrical signal, and z/r_{0z}^2 is again the falloff in amplitude with distance, including the obliquity factor z/r_{0z}. The signal due to the reflecting point is thus derived by integrating over the transducer area $s(x_0, y_0)$. Performing the temporal convolution $e_h(t)$ becomes

$$e_h(x_z, y_z, t) = \int\int \left[\int\int s(x_0, y_0) \exp{(ikr_{0z})}p\left(t - \frac{r_{0z}}{c} - \frac{r'_{0z}}{c}\right)\left(\frac{z}{r_{0z}^2}\right)dx_0 dy_0\right]$$

$$\times s(x'_0, y'_0) \exp{(ikr'_{0z})}\left(\frac{z}{r'^2_{0z}}\right)dx'_0 dy'_0 e^{-i\omega_0 t} * a(t) * b(t) \quad (9.19)$$

where the primed coordinates are used to distinguish the integration of the reflected components. The uncompensated envelope response $e(t)$ for a general object with reflectivity $R(x, y, z)$ having a uniform attenuation α is then given by

$$e(t) = \left|\int\int\int e^{-2\alpha z}R(x, y, z)e_h(x, y, t)dx dy dz\right|. \quad (9.20)$$

STEADY-STATE APPROXIMATIONS TO THE DIFFRACTION FORMULATION

In equations (9.17) and (9.19) the pulse envelope $p(t)$ appears in the spatial integral defining the system spatial response. Physically, the transmitted pulse from different portions of the transducer arrives at each reflecting point at different times, and the reflected pulse arrives at different portions of the transducer at different times. If the pulse is relatively short, the resultant transducer output will depend on which portion is being excited. For very long pulse envelopes, however, the entire transducer is excited by essentially the same sinusoidal waveform so that a steady-state analysis represents a good approximation.

In making the steady-state approximation, we assume that

$$p\left(t - \frac{r_{0z}}{c}\right) \simeq p\left(t - \frac{z}{c}\right) \quad (9.21)$$

which makes $p(\cdot)$ independent of x and y, and allows it to be moved outside the integral in equations (9.17) and (9.19), where it affects only temporal or depth resolution, not lateral resolution. Physically, we are assuming that the envelope of the transmitted waveform from all parts of the transducer arrives at each depth plane z at approximately the same time. Similarly, the waveforms reflected off each reflecting point arrive at all portions of the transducer at approximately the same time. Thus the lateral resolution considerations are governed solely by the relative phases of the received waveforms at different portions of the transducer, not by the envelope $p(t)$.

To estimate the validity of this steady-state approximation, we can calculate the maximum delay difference between the center and edge of a transducer of extent D for a reflecting point in the center of the beam. For the steady-state approximation to hold, the duration τ of the pulse envelope $p(t)$ should be long compared to this delay difference. This ensures that the entire transducer is simultaneously insonified for a reasonable duration. Thus the approximation criterion may be stated as

$$\tau \gg \frac{r_{0z_{max}} - z}{c} \tag{9.22}$$

where $r_{0z_{max}} = \sqrt{(D/2)^2 + z^2}$. Since relatively narrow angles are usually involved, we can use the approximation $r_{0z_{max}} \simeq z + (D/2)^2/2z$, providing a criterion for the steady-state approximation given by

$$\tau \gg \frac{D^2}{8zc}. \tag{9.23}$$

For a transducer size of $D = 2$ cm and a depth of $z = 10$ cm, the pulse duration must be significantly greater than 0.3 μsec. This is achieved by most systems since durations of about 1.0 μsec are typical. However for larger transducers or shorter depths, this steady-state approximation will be poor. A more accurate analysis will be considered subsequently.

Applying the steady-state approximation of (9.21), the transmitted field amplitude at depth z is given by

$$u(x_z, y_z, t) = \left[\int \int s(x_0, y_0) \exp{(ikr_{0z})} \left(\frac{z}{r_{0z}^2}\right) dx_0 dy_0 \right]$$
$$\times p\left(t - \frac{z}{c}\right) \exp{(-i\omega_0 t)} * a(t) \tag{9.24}$$

where the lateral and temporal or depth resolution functions are clearly separated. The overall round-trip response to a point reflector $e_h(x_z, y_z, t)$ using equation (9.19) then becomes

$$e_h(t) = \left[\int \int s(x_0, y_0) \exp{ikr_{0z}} \left(\frac{z}{r_{0z}^2}\right) dx_0 dy_0 \right]^2 e^{-i\omega_0 t} \tilde{p}\left(t - \frac{2z}{c}\right) \tag{9.25}$$

where $\tilde{p}(t)$ is the effective pulse envelope considering all the linear processes where

$$\tilde{p}\left(t - \frac{2z}{c}\right) e^{-i\omega_0 t} = p\left(t - \frac{2z}{c}\right) e^{-i\omega_0 t} * a(t) * b(t). \tag{9.26}$$

The envelope-detected output of the system using a generalized object and compensated for tissue attenuation is then given by

$$e_c(t) = K \left| \int \int \int R(x, y, z) \left[\int \int s(x_0, y_0) e^{ikr_{0z}} \left(\frac{z}{r_{0z}^2}\right) dx_0 dy_0 \right]^2 \tilde{p}\left(t - \frac{2z}{c}\right) dx\,dy\,dz \right|. \tag{9.27}$$

Note in equations (9.25) and (9.27) that the integration defining the lateral resolution is squared because of the symmetry of the transmitting and receiving operation. Thus, in the steady-state approximation, the overall lateral resolution is the product of the identical transmitter and receiver patterns. We will now proceed to study approximations to the lateral response function in the brackets so as to make the resultant expressions more tractable and subject to analysis. We separate out the lateral response of the transmit and receive operations as given by

$$h(x_z, y_z) = \int \int s(x_0, y_0) \exp{(ikr_{0z})}\left(\frac{z}{r_{0z}^2}\right) dx_0 dy_0. \qquad (9.28)$$

As a first approximation we apply the paraxial approximation where the z/r_{0z}^2 term becomes $1/z$ since $z \simeq r_{0z}$ near the axis. It should be emphasized that this approximation is relatively insensitive because of the multiplicative effect of this term. This is in sharp distinction to the r_{0z} term in the exponent because of its greater sensitivity. The approximations of r_{0z} in the exponent are divided into the Fresnel and Fraunhofer regions, often referred to as the near-field and far-field regions.

FRESNEL APPROXIMATION

In the Fresnel region [Goodman, 1968] we approximate r_{0z} in the exponent as the first two terms of the binomial expansion of

$$r_{0z} = z\sqrt{1 + \frac{(x_0 - x_z)^2 + (y_0 - y_z)^2}{z^2}} \qquad (9.29)$$

as given by

$$r_{0z} \simeq z + \frac{(x_0 - x_z)^2 + (y_0 - y_z)^2}{2z}. \qquad (9.30)$$

This approximation is valid in regions where $z^3 \gg (\pi/4\lambda)[(x_0 - x_z)^2 + (y_0 - y_z)^2]_{max}^2$. This inequality ensures that the exponent due to the third term in the binomial expansion is significantly less than unity. As in optics, this approximation is quite accurate for systems with reasonable angular fields. The lateral response at plane z with the Fresnel approximation becomes

$$h(x_z, y_z) = h(x, y, z) = e^{ikz}\frac{s(x, y)}{z} ** \exp\left[i\frac{k}{2z}(x^2 + y^2)\right] \qquad (9.31)$$

where the convolution formulation and the generalized x, y coordinates have been used for convenience. The phase factor $\exp{(ikz)}$ represents the phase shift at different depths. The convenient convolution form can be used since, using the Fresnel approximation, the impulse response is space invariant, depending solely on the difference between the spatial coordinates.

This impulse response can be used to find the field amplitude $h(x_z, y_z)$ at any plane z due to any transducer distribution $s(x, y)$. The effect of this convolution is relatively complex. The amplitude distribution of a circular disk transducer is shown in Fig. 9.7. As is seen, the field amplitude in the immediate vicinity of the transducer has an oscillatory pattern whose extent is approximately a geometric extension of the transducer. At a depth of $D^2/4\lambda$, the oscillatory pattern diminishes and the pattern begins to diverge uniformly. This distance is referred to as the near field by some authors. The oscillatory pattern extends to a depth of about $D^2/2\lambda$ for a square transducer, twice that of the circular transducer. At a distance of D^2/λ the 3-db width of the beam equals the transducer diameter. Beyond this depth we are clearly in the Fraunhofer or far-field pattern, which is the next subject of discussion. In this region the shape of the pattern remains fixed and its size linearly increases with depth, representing a fixed angular pattern having an angle of approximately λ/D.

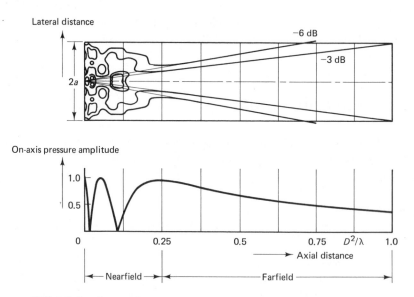

FIG. 9.7 Steady-state field pattern of a circular transducer. (Courtesy of Siemens Gammasonics, Inc., Vetrasound Division.)

Having described the steady-state diffraction behavior in the Fresnel region, we can return to defining the received signal from a reflecting object as in equation (9.21). In the Fresnel region the envelope detected signal is given by

$$e(t) = K \left| \int \int \int \frac{\exp{(-2\alpha z)}}{z^2} R(x, y, z) e^{i2kz} \right.$$
$$\left. \times \left\{ s(x, y) ** \exp\left[i\frac{k}{2z}(x^2 + y^2) \right] \right\}^2 \tilde{p}\left(t - \frac{2z}{c} \right) dx\,dy\,dz \right|. \quad (9.32)$$

The overall attenuation factor is $\exp{(-2\alpha z)}/z^2$ due to the tissue attenuation α

and the diffraction spreading in both directions, $1/z^2$. As before, we can assume that this attenuation factor varies very slowly compared to the effective pulse envelope $\tilde{p}(t - 2z/c)$. Thus the envelope function acts as a delta function on the attenuation factor, allowing it to be taken outside the integral as a time variation $e^{-\alpha ct}/(ct/2)^2$. As in equation (9.3), a compensating gain variation is used where $g(t) = (ct)^2 e^{\alpha ct}$, resulting in a compensated output signal given by

$$e_c(t) = K \left| \int\int\int R(x, y, z)e^{i2kz} \left\{ s(x, y) ** \exp\left[i\frac{k}{2z}(x^2 + y^2)\right] \right\}^2 \right.$$
$$\left. \times \tilde{p}\left(t - \frac{2z}{c}\right) dx\,dy\,dz \right|. \qquad (9.33)$$

For an A scan, using a stationary transducer positioned at $x = 0$, $y = 0$, the resulting estimate of the reflection coefficient along the z direction is given by

$$\hat{R}(0, 0, z) = K \left| R(x, y, z)e^{i2kz} *** \left\{ s(x, y) ** \exp\left[i\frac{k}{2z}(x^2 + y^2)\right] \right\}^2 \tilde{p}\left(\frac{2z}{c}\right) \right|$$
$$\text{evaluated at } x = 0, y = 0. \quad (9.34)$$

For a B scan, again with the transducer moving in the x direction along the $y = y_0$ line with a velocity v the gain-compensated signal is approximated as

$$e_c(t) = K \sum_{n=0}^{N} \left| \int\int\int R(x, y, z)e^{i2kz} \right.$$
$$\left. \times \left\{ s(x - vnT, y - y_0) ** \exp\left[i\frac{k}{2z}(x^2 + y^2)\right] \right\}^2 \tilde{p}\left(t - \frac{2z}{c}\right) dx\,dy\,dz \right|. \quad (9.35)$$

The resultant displayed estimate of the reflectivity in the $y = y_0$ plane is then given by

$$\hat{R}(x, y_0, z) = K \sum_{n=0}^{N} \left| R(x, y, z)e^{i2kz} *** \left\{ s(x, y) ** \exp\left[i\frac{k}{2z}(x^2 + y^2)\right] \right\}^2 \right.$$
$$\left. \times \tilde{p}\left(\frac{2z}{c}\right) \right| \qquad \text{evaluated at } x = vnT, y = y_0 \quad (9.36)$$

where T, as before, is the time of each scan line.

It is instructive to consider the affect of the phase-shift term $\exp(i2kz)$ representing the round-trip time to each plane. If we have reflections at one plane only, each reflection will experience the same phase shift. This phase term, $\exp(i2kz)$, will then disappear when the magnitude is taken, representing envelope detection. However, given reflections at different depths, each reflection will be associated with a different phase shift. The resultant signals, as represented in equation (9.35), will add constructively or destructively, depending on their relative phases. This addition and cancellation of signals results in a mottled pattern known as "speckle" whose properties are studied later in this chapter. If this phase factor is ignored, the resultant image will be the incoherent average of the reflectivity, devoid of speckle.

FRAUNHOFER APPROXIMATION

The general results in the Fresnel region, under the approximations outlined, are valid at all depths. However, in the far-field or Fraunhofer region, the expressions can be simplified. This not only provides simpler mathematical operations, but also represents considerable insight into the nature of the imaging patterns.

We return to the convolution relationship of equation (9.31) in integral form representing the lateral impulse response as given by

$$h(x_z, y_z) = \frac{e^{ikz}}{z} \int \int s(x_0, y_0) \exp \left\{ i \frac{k}{2z} [(x_z - x_0)^2 + (y_z - y_0)^2] \right\} dx_0 dy_0. \quad (9.37)$$

This equation can be restructured as

$$h(x_z, y_z) = \frac{\exp[ik(r_z^2/2z + z)]}{z} \int \int s(x_0, y_0) \exp \left(\frac{ikr_0^2}{2z} \right)$$

$$\times \exp \left[-i \left(\frac{2\pi}{\lambda z} \right) (x_0 x_z + y_0 y_z) \right] dx_0 dy_0 \quad (9.38)$$

where $r_z^2 = x_z^2 + y_z^2$ and $r_0^2 = x_0^2 + y_0^2$. As before, the phase factor outside the integral can be ignored when investigating a specific plane. However, for reflections at various depths this z-dependent phase shift results in coherent speckle. As for the quadratic phase factor within the integral, for a transducer having a maximum lateral dimension of D, the exponent will have a maximum value of $\pi D^2/4\lambda z$, or approximately $D^2/\lambda z$. Thus for depths greater than D^2/λ, the exponent will be less than 1 radian, and can be neglected. This leaves only a two-dimensional Fourier transform kernel in the integral, so that (9.38) can be approximated as

$$h(x_z, y_z) \simeq \frac{e^{iv}}{z} \int \int s(x_0, y_0) \exp \left[-i \left(\frac{2\pi}{\lambda z} \right) (x_0 x_z + y_0 y_z) \right] dx_0 dy_0$$

$$\simeq \frac{e^{iv}}{z} \mathcal{F}\{s(x_0, y_0)\} \qquad u = \frac{x_z}{\lambda z} \qquad v = \frac{y_z}{\lambda z} \quad (9.39)$$

where $v = k(r_z^2/2z + z)$ and $\mathcal{F}$ is the Fourier transform operator using spatial frequency coordinates u and v as indicated. Thus in the far-field or Fraunhofer region, where $z \gg D^2/\lambda$, the compensated received signal is given by

$$e_c(t) = K \left| \int \int \int R(x, y, z) e^{i2v} [\mathcal{F}\{s(x, y)\}]^2 \tilde{p} \left(t - \frac{2z}{c} \right) dx\, dy\, dz \right|. \quad (9.40)$$

In this far-field region the received signals and reflectivity estimates can be obtained by substituting for the lateral response terms $\{s(x, y) ** \exp[i(k/2z)(x^2 + y^2)]\}^2$ in equations (9.32) through (9.36) the simpler expression $[\mathcal{F}\{s(x, y)\}]^2$. This formulation serves to illustrate the performance problems of collimated ultrasonic imaging systems in the far field. This far-field response, the Fourier transform of the source function, is known as the "diffraction-limited" response since it represents the best resolution for a given source configuration.

Consider a square transducer where $s(x, y) = \text{rect}\,(x/D)\,\text{rect}\,(y/D)$. The resultant lateral spatial pattern in the far field is given by

$$h(x_z, y_z) = \frac{e^{iv}}{z}\,\mathcal{F}\left\{\text{rect}\left(\frac{x_0}{D}\right)\text{rect}\left(\frac{y_0}{D}\right)\right\} = \frac{e^{iv}D^2}{z}\,\text{sinc}\left(\frac{Dx_z}{\lambda z}\right)\text{sinc}\left(\frac{Dy_z}{\lambda z}\right) \quad (9.41)$$

where $\text{sinc}\ x = \sin(\pi x)/\pi x$. The effective width of the sinc function may be defined as the region where its argument is between $\pm\frac{1}{2}$. In this region the beam width at any depth z is $\lambda z/D$. An approximation to the total field is shown in Fig. 9.8, showing the near collimated behavior to a depth D^2/λ and then the diverging behavior in the far field. The various perturbations within the Fresnel region shown in Fig. 9.7 have been ignored. As is shown, a large transducer, having a maximum diameter of D_1, remains approximately collimated to a depth D_1^2/λ. If a smaller transducer of dimension D_2 is used in an attempt to improve lateral resolution, its performance rapidly deteriorates at greater depths because of the reduced value of D_2^2/λ.

Figure 9.8 illustrates the difficulty in designing an ultrasonic transducer. One approach is to provide approximately near-field performance at all depths.

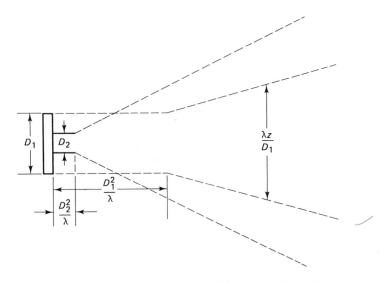

FIG. 9.8 Approximate field patterns for different transducer sizes.

In that case the transducer dimension D is chosen such that the limit of the near-field region, D^2/λ, is equal to the maximum depth $z_{\max}$. Thus for collimated imaging the optimum transducer size becomes

$$D_{\text{opt}} \simeq \sqrt{\lambda z_{\max}}. \quad (9.42)$$

For a maximum depth of 20 cm and a wavelength of 0.5 mm this represents a transducer size of 1.0 cm. This is a representative figure for modern instruments. Improvements in resolution can be obtained by reducing the wavelength. However, since the attenuation α is frequency dependent, the increased frequency results in excessive attenuation at the greater depths. The situation is quite different, however, in systems designed for superficial imaging of the eyes, thyroid, vessels in the neck, and so on. These systems involve a maximum depth of about 4 cm. Because of the reduced attenuation problem frequencies as high as 10 MHz ($\lambda = 0.15$ mm) can be used. In these cases the optimum transducer size D is 2.4 mm, representing a considerable improvement in resolution.

ACOUSTIC FOCUSING

One method of modifying the performance shown in Figs. 9.7 and 9.8 is through the use of acoustic focusing, as illustrated by the lens system of Fig. 9.9 [Goodman, 1968]. Here an acoustic lens is used to obtain optimum resolution at a

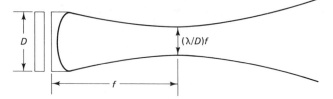

FIG. 9.9 Acoustic focusing system.

single depth plane and its vicinity. The acoustic lens is generally made of a plastic material which has a velocity of propagation *greater* than that of water. Thus, unlike the case of light optics, the refractive index n is less than unity, where unity represents the index of the surrounding water. The relative phase delay through the material at each lateral position x, y is given by

$$\theta(x, y) = k(n - 1)d(x, y) \tag{9.43}$$

where $d(x, y)$ is the thickness of the lens at each position. We can approximate the spherical surface as a quadratic surface in the paraxial region. In that case $d(x, y)$ can be approximated by

$$d(x, y) \simeq K + \frac{x^2 + y^2}{2R} \tag{9.44}$$

where R is the radius of curvature of the lens and K is a constant. Ignoring constant phase factors the phase shift of the lens is thus given by

$$\theta(x, y) = k(n - 1)\frac{x^2 + y^2}{2R} = -k\frac{x^2 + y^2}{2f} \tag{9.45}$$

where f, the focal length is $-(n - 1)/R$. Since the refractive index is less than one, this system is a positive converging lens.

To study the field patterns this phase factor is added to the source distribution in the Fresnel equation (9.38) as given by

$$h(x_z, y_z) = \frac{e^{iv}}{z} \int \int s(x_0, y_0) \exp\left(-i\frac{k}{2f}r_0^2\right) \exp\left(i\frac{k}{2z}r_0^2\right)$$

$$\times \exp\left[-i\left(\frac{2\pi}{\lambda z}\right)(x_0 x_z + y_0 y_z)\right] dx_0 dy_0. \tag{9.46}$$

As can be seen, where $z = f$, at a depth equal to the focal length, the field amplitude is the Fourier transform of the source distribution. As before, this results in an effective lateral beam width, at the focal plane, of $\lambda f / D$. Thus a large aperture system can result in a relatively small, well-defined beam at this depth. At depths outside this region the quadratic phase factor returns, resulting in various diverging patterns. For "weakly focused systems" where the system F number f/D is appreciably greater than 1, the behavior is similar to that shown in Fig. 9.9 where the resolution is best at the focal plane and degrades gradually on either side.

WIDEBAND DIFFRACTION

The previous formulations all used the steady-state approximation where the pulse envelope in equation (9.17), $p(t - r_{0z}/c)$, is approximated as $p(t - z/c)$, as in equation (9.24). As previously indicated, this approximation can be poor in systems with short pulses and/or large apertures. To study a more exact solution [Norton, 1976] for the short pulse or wideband case we will consider the transmitted field pattern at a depth z. We simplify the formulation of (9.17) as

$$u(x_z, y_z, t) = \int \int s(x_0, y_0) \exp\left(ikr_{0z}\right)\tilde{p}\left(t - \frac{r_{0z}}{c}\right)$$

$$\times \left(\frac{z}{r_0^2}\right) dx_0 dy_0 \exp\left(-i\omega_0 t\right) \tag{9.47}$$

where

$$\tilde{p}\left(t - \frac{r_{0z}}{c}\right) \exp\left(-i\omega_0 t\right) = p\left(t - \frac{r_{0z}}{c}\right) \exp\left(-i\omega_0 t\right) * a(t)$$

and $\tilde{p}(t)$ is the effective pulse envelope in the transmitted waveform.

We again use the Fresnel approximation of equation (9.30) since the same geometric constraints apply to the wideband case. The amplitude factor z/r_{0z}^2 again becomes $1/z$ and the r_{0z} in the exponent and in the argument of $\tilde{p}$ are replaced by the first two terms of the binomial expansion, giving

$$u(x_z, y_z, t) = \frac{e^{ikz}}{z} \int \int s(x_0, y_0) \exp\left\{i\frac{k}{2z}[(x_z - x_0)^2 + (y_z - y_0)^2]\right\}$$

$$\times \tilde{p}\left[t - \frac{z}{c} - \frac{(x_z - x_0)^2}{2zc} - \frac{(y_z - y_0)^2}{2zc}\right] dx_0 dy_0. \tag{9.48}$$

Expanding the quadratic terms, we obtain the two-dimensional transform relationship

$$u(x_z, y_z, t) = \frac{e^{iv}}{z} \mathcal{F} \left\{ s(x_0, y_0) \exp\left(\frac{ikr_0^2}{2z}\right) p'' \left[t - \frac{z}{c} - \frac{(x_z - x_0)^2}{2zc} - \frac{(y_z - y_0)^2}{2zc} \right] \right\}$$

$$\times \exp(-i\omega_0 t), \qquad u = \frac{x_z}{\lambda z}, \quad v = \frac{y_z}{\lambda z}. \qquad (9.49)$$

As with the Fresnel integral in the steady-state case, (9.49) is difficult to evaluate. We can again simplify the formulation in the far-field region where quadratic terms in x_0, y_0 in both the exponent and the argument of $\tilde{p}$ can be neglected. In addition, we restrict our formulation to the x dimension to avoid excessive complexity in the notation. In that case the transmitted field is given by

$$u(x_z, t) \simeq \frac{e^{iv}}{z} \mathcal{F} \left\{ s(x_0) p'' \left(t - T + \frac{x_z x_0}{zc} \right) \right\} \exp(-i\omega_0 t) \qquad (9.50)$$

where $T = z/c + x_z^2/2zc$, representing the approximate time delay from the origin to the reflecting point.

As an illustrative example, assume a rectangular source and a rectangular pulse envelope where $s(x_0) = \text{rect}(x_0/D)$ and $\tilde{p}(t) = \text{rect}(t/\tau)$. The field amplitude is then given by

$$u(x_z, t) = \frac{\exp[-i(\omega_0 t - v)]}{z} \mathcal{F} \left\{ \text{rect}\left(\frac{x_0}{D}\right) \text{rect}\left(\frac{x_0 - (t - T)(zc/x_z)}{\tau zc/x_z}\right) \right\}. \qquad (9.51)$$

This product provides two general options: the cases where $D > |\tau zc/x_z|$ and where $D < |\tau zc/x_z|$. In each case the smaller rectangle dominates the product of the two functions. Within each option we have three regions, corresponding to different times, where the smaller rectangular function either fits into the larger one or is at either end where the resultant area is reduced. Thus each option is represented by the sum of three functions, each being a Fourier transform of a rectangle multiplied by rectangular time functions which define the time intervals in which the different regions occur. Since the time response represents our depth resolution, it is convenient to define $t' = (T - t)c$, representing time as an equivalent depth in terms of an effective retarded time. Similarly, we let $\tau' = \tau c$, the equivalent length of the pulse. We also define $\phi = x_z/z$ as a close approximation to the angle each object point makes with the origin.

For simplification, to avoid carrying phase factors, we limit our expression to $|u(x_z, t)| = |u(\phi, t')|$. This is in keeping with the remainder of this chapter, where phase factors have been dropped where appropriate since envelope detection is assumed. The overall response is therefore given by

$$|u(\phi, t')| = \frac{1}{z} \left\{ D \left| \text{sinc}\left(\frac{D\phi}{\lambda}\right) \right| \text{rect}\left(\frac{t'}{\tau' - D\phi}\right) + \frac{\lambda}{\pi\phi} \left| \sin\left[\frac{1}{\lambda}\left(t' + \frac{\tau'}{2} + \frac{D\phi}{2}\right)\right] \right| \right.$$

$$\left. \times \text{rect}\left(\frac{t' + \tau/2}{D\phi}\right) + \frac{\lambda}{\pi\phi} \left| \sin\left[\frac{1}{\lambda}\left(t' - \frac{\tau'}{2} - \frac{D\phi}{2}\right)\right] \right| \text{rect}\left(\frac{t' - \tau/2}{D\phi}\right) \right\} \qquad (9.52)$$

for $\tau' > D\,|\phi|$ and

$$|u(\phi, t')| = \frac{1}{z}\left\{\left|\frac{\tau'}{\phi}\right|\text{sinc}\left(\frac{\tau'}{\lambda}\right)\left|\text{rect}\left(\frac{t'}{D\phi - \tau'}\right) + \frac{\lambda}{\pi\phi}\left|\sin\left[\frac{1}{\lambda}\left(t' + \frac{\tau'}{2} + \frac{D\phi}{2}\right)\right]\right|\right.\right.$$

$$\left.\times \text{rect}\left(\frac{t' + D\phi/2}{\tau'}\right) + \frac{\lambda}{\pi\phi}\left|\sin\left[\frac{1}{\lambda}\left(t' - \frac{\tau'}{2} + \frac{D\phi}{2}\right)\right]\right|\text{rect}\left(\frac{t' - D\phi/2}{\tau'}\right)\right\} \quad (9.53)$$

for $\tau' < D\,|\phi|$. In this case the absolute value of the sum becomes the sum of the absolute values because the terms occur at different times. Notice that in the second and third terms of each expression it becomes more convenient to express the transform as a sine rather than a sinc function.

These relatively complex equations show the interaction between the temporal or depth response, using the coordinate t', and the lateral or angular response, using the coordinate ϕ. We can study some special cases in the interest of clarification. Setting t' equal to zero, we limit ourselves to the depth plane $z = Tc$ and obtain the lateral response as

$$|u(\phi, 0)| = \frac{D}{z}\left|\text{sinc}\left(\frac{D\phi}{\lambda}\right)\right|. \quad (9.54)$$

Similarly, we obtain the response along the z axis by setting $\phi = 0$ and obtaining

$$|u(0, t')| = \frac{D}{z}\text{rect}\left(\frac{t'}{\tau'}\right). \quad (9.55)$$

On the t' and ϕ axes we observe the relatively straightforward responses that were used in the steady-state analysis. In the steady-state analysis these were also the off-axis responses. If we plot the effective extent of the response in space, for example a plot of ϕ versus t' indicating the lateral and depth resolution, it would be a rectangle under the steady-state approximation since the temporal and spatial responses are assumed to be noninteracting.

A typical response using the more accurate formulation of equation (9.52) is shown in Fig. 9.10 for a relatively wide pulse. This figure can only show the approximate extent of the response since the amplitude at each value of ϕ and t' requires the third dimension. It does illustrate the important deviations from the steady-state approximation. We see the diagonal arms resulting primarily from the last two terms in equation (9.52). These fall off as $1/\phi$, which is not shown in the figure. These represent undesired responses well removed from the region of interest. As the pulse length τ increases, the relative amount of energy in the diagonal arms decreases to the point where the effective extent of the response approaches a rectangle, as in the steady-state approximation.

In the other extreme, for very short pulses, the response is dominated by the first term in (9.53), as given by

$$|u(\phi, t')| \simeq \frac{1}{z}\left(\frac{\tau'}{\phi}\right)\text{rect}\left(\frac{t'}{D\phi}\right) \quad (9.56)$$

which is plotted in Fig. 9.11. Here the response is seen to be a very short pulse, or delta function, for $\phi = 0$ and increasing in width as ϕ increases. The shading

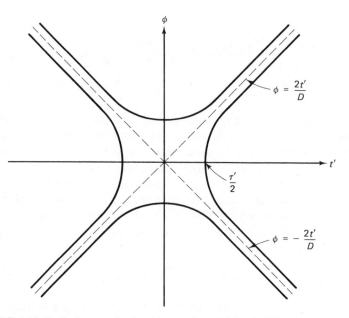

FIG. 9.10 Typical transmitted pattern, using wideband diffraction considerations.

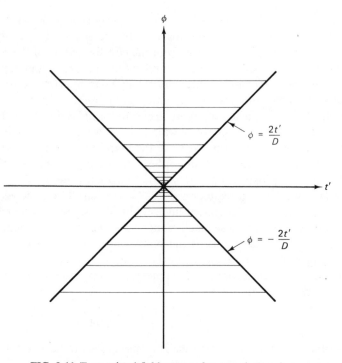

FIG. 9.11 Transmitted field pattern for very short pulses.

indicates the falloff in amplitude with ϕ. As can be seen from Figs. 9.10 and 9.11, the pulse waveform should be chosen to minimize the volume in the $t'-\phi$ space, thus providing the best compromise between depth and lateral resolution.

Using the same techniques, the complete transmitter pattern $u(x_z, y_z, t)$ can be studied. The round-trip response to a unity reflecting point $e_h(x_z, y_z, t)$ is again derived using equation (9.17). The received signal from a generalized reflecting object $R(x, y, z)$ is derived using equation (9.20). In each of these equations the far-field and paraxial approximations can again be applied to simplify the computations and achieve insightful results.

ULTRASONIC CHARACTERISTICS OF TISSUE

Attenuation

We have thus far assumed a uniform attenuation coefficient throughout the object being studied. However, in practice, the attenuation coefficient is a function of both the particular tissue and of the frequency of the propagating wave [Woodcock, 1979]. The attenuation mechanisms in biological materials are not well understood. In many common fluids, such as water, the attenuation is primarily due to viscous absorption. In these cases the attenuation is proportional to the square of the frequency. In most biological materials, however, in the frequency range 1.0 to 10 MHz, the attenuation varies directly with the frequency. This mechanism is usually attributed to a relaxation process in which energy is removed from the propagating wave by an oscillating particle and then returned at a later time. Some typical values for attenuation at 1.0 MHz are given in Table 9.1.

TABLE 9.1

ATTENUATION AT 1.0 MHz

Material	Attenuation Coefficient (db/cm)
Air	10
Blood	0.18
Bone	3–10
Lung	40
Muscle	1.65–1.75
Other soft tissues	1.35–1.68
Water	0.002

As indicated, the coefficient of the biological tissues varies approximately directly with frequency while those of water and air vary as the square of the frequency. In studies where the propagation path is primarily soft-tissue structures of comparable attenuation, such as the abdomen, a fixed-gain compensation system is usually adequate. In other cases involving blood pools or fluid regions, it is often desirable to vary the exponent of the gain compensation over the path. Many commercial scanners make this option available.

Velocity

The basis for ultrasonic imaging in the reflection mode is the assumption of a constant propagation velocity throughout the body. The round-trip time of each echo is used to determine its depth. Fortunately, although various materials exhibit profound changes in their acoustic velocity, the soft tissues of the body are limited to a range of about $\pm 5\%$. Some representative values are given in Table 9.2.

TABLE 9.2

PROPAGATION VELOCITY

Tissue	Mean Velocity (m/sec)
Air	330
Fat	1450
Aqueous humor of eye	1500
Vitreous humor of eye	1520
Human tissue, mean value	1540
Brain	1541
Liver	1549
Kidney	1561
Spleen	1566
Blood	1570
Muscle	1585
Lens of eye	1620
Skull bone	4080
Water	1480

Variations in velocity cause small geometric distortions in the reproduced images since the display system assumes a constant propagation velocity. In addition, velocity variations can distort and deflect the propagating beam, causing additional geometric errors. In some cases it has been found that the distribution of propagation velocity throughout the object of interest, $c(x, y)$, contains clinically useful information. For example, certain malignant tumors exhibit increased propagation velocity. A method for reconstructing both $c(x, y)$ and $\alpha(x, y)$, using the techniques of computerized tomography, is shown in Chapter 11.

Reflectivity

Our development assumes that the reflectivity of tissues can be modeled as an array of weakly reflecting isotropic scatterers having a reflectivity $R(x, y, z)$. Because of the isotropic nature of the reflections, a falloff of $1/z$ was assumed as in equation (9.1). In general, the reflectivity depends on both the shape and the material in a relatively complex manner [Nicholas, 1977]. The simplest behavior occurs at a planar interface between two materials. The resultant reflection is called a *specular reflection*, as differentiated from the diffuse reflections we have assumed. The planar surface acts as a mirror and reflects the wave at an angle equal and opposite to the angle of incidence. In this case the amplitude of the reflected wave received by the transducer becomes a strong function of the position and angle of the transducer. The mathematical development for specular reflectors of general shapes is more complex than that given for isotropic scatterers. However, the general concepts of resolution and diffraction considerations remain the same.

In the earlier ultrasonic instruments, where binary images were produced which essentially outlined organs and lesions, these specular reflections were the most significant information. In modern instrumentation with large dynamic range and gray-scale displays, the diffusely reflecting, isotropically scattering echoes have become the most significant, hence the use of the model in this chapter.

The reflectivity is dependent on changes in the acoustic impedance. This impedance Z relates the pressure P to the particle velocity v as given by

$$P = Zv \qquad (9.57)$$

where the acoustic impedance is given by

$$Z = \rho c \qquad (9.58)$$

where ρ is the density and c the velocity.

The behavior at a planar interface between two materials can be studied using Fig. 9.12. For equilibrium, the total pressure on each side of the interface must be equal, and the particle velocity on each side of the interface must be continuous. These conditions are satisfied when

$$P_i + P_r = P_t \qquad (9.59)$$

and

$$v_i \cos \theta_i - v_r \cos \theta_r = v_t \cos \theta_t \qquad (9.60)$$

where the subscripts i, r, and t indicate incident, reflected, and transmitted components, respectively. Using Snell's law, we have

$$\frac{\sin \theta_i}{\sin \theta_t} = \frac{c_1}{c_2} \tag{9.61}$$

and, as in electromagnetic theory, we set the angle of incidence θ_i equal to the angle of reflectance θ_r.

The reflectivity, defined as the ratio of the reflected pressure to the incident pressure, is found using equations (9.57) through (9.61) and is given by

$$R = \frac{P_r}{P_i} = \frac{Z_2 \cos \theta_i - Z_1 \cos \theta_t}{Z_2 \cos \theta_i + Z_1 \cos \theta_t} \tag{9.62}$$

where Z_1 and Z_2 are the acoustic impedances in the two interfacing media. At normal incidence where $\theta_i = \theta_t = 0$, we have

$$R = \frac{Z_2 - Z_1}{Z_2 + Z_1}. \tag{9.63}$$

Table 9.3 gives the reflectivity R at normal incidence for a variety of tissue interfaces.

TABLE 9.3

REFLECTIVITY OF NORMALLY INCIDENT WAVES

Materials at Interface	Reflectivity
Brain–skull bone	0.66
Fat–bone	0.69
Fat–blood	0.08
Fat–kidney	0.08
Fat–muscle	0.10
Fat–liver	0.09
Lens–aqueous humor	0.10
Lens–vitreous humor	0.09
Muscle–blood	0.03
Muscle–kidney	0.03
Muscle–liver	0.01
Soft tissue (mean value)–water	0.05
Soft tissue–air	0.9995
Soft tissue–PZT5 crystal	0.89

Here we see that the interfaces between soft tissues have a reflectivity of under 0.10, representing less than 1 % of the energy being reflected. This coincides with our "weakly reflecting" assumption, where multiple reflections were ignored. However, a number of other interfaces, such as between tissue and bone, tissue and air, and tissue and the transducer, have strong reflections. Thus certain clinical situations can result in multiple reverberations, giving rise to false echoes.

COMPOUND SCAN FOR SPECULAR INTERFACES

At specular interfaces, as shown in Fig. 9.12, the wave is reflected in mirror-like fashion where $\theta_r = -\theta_i$. As indicated in Fig. 9.13, this can result in the specular reflection from an organ interface completely missing the transducer. The resultant image will therefore be missing some of the information relating to the interface.

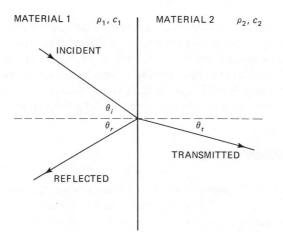

FIG. 9.12 Behavior at a plane surface.

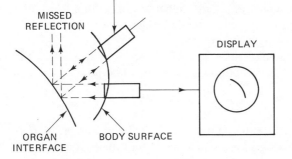

FIG. 9.13 Effect of a compound B scan.

This problem can be alleviated, with manual transducers, by utilizing a compound translation-rocking motion [Woodcock, 1979]. At each lateral position a number of line scans are obtained with the transducer angulated in different directions, all in the plane of the desired cross section. Potentiometers record the position and angle of the transducer so that the received echoes at each angle can be appropriately recorded in the display. Thus the transducer eventually reaches a position and angle, as shown in Fig. 9.13, where the beam is perpendicular to the interface so that the specular reflection is received. The use of higher frequencies has increased the ratio of diffuse to specular echoes since the amplitude of the diffuse echoes increases as the square of the frequency. However, these are still considerably lower in amplitude than the specular echoes so that a large dynamic range is required to preserve them. Following detection, most ultrasonic systems use nonlinear compression, prior to display, to compress the large specular echoes and enhance the weaker diffuse echoes.

The array systems, discussed in the next chapter, use stationary transducer arrays and thus do not involve compound scanning. However, if a large array subtends a relatively large angle with the region of interest, it reduces its angular sensitivity. One mechanical commercial scanner uses eight rotating transducers immersed in a water bath. These separate views are added to provide the effect of a compound scan and minimize the angular sensitivity of specular interfaces.

NOISE CONSIDERATIONS

Unlike x-ray, with its signal-dependent Poisson noise, the noise in ultrasonic systems is governed by additive Gaussian noise resulting from the transducer and the first amplifier. The resultant signal-to-noise ratio is therefore the ratio of the received signal power at the transducer terminals to the average noise power $\overline{e_n^2}$.

In estimating the signal, it must be emphasized that the attenuation compensation, defined in equation (9.3), is performed beyond the transducer and thus does not generally affect the signal-to-noise ratio. Thus, as would be expected, reflections from greater depths, which experience increased attenuation, result in a reduced signal-to-noise ratio.

The signal-to-noise ratio at each depth z_0 is defined as the peak signal power received at that depth divided by the noise power as given by

$$\text{SNR} = \frac{E_0^2}{\overline{e_n^2}} \tag{9.64}$$

where E_0 is the peak value of $e_0(t)$, the signal envelope $e(t)$ derived from depth plane $z = z_0$. This signal $e_0(t)$, in a single transducer system, using steady-state diffraction theory, from equation (9.32) is given by

$$e_0(t) = K \left| \frac{e^{-2\alpha z_0}}{z_0^2} \tilde{p}\left(t - \frac{2z_0}{c}\right) \int \int R(x, y, z_0)[s(x, y) ** e^{i(kr^2/2z_0)}]^2 dxdy \right|. \tag{9.65}$$

The peak value E_0 is given by

$$E_0 = K \frac{e^{-2\alpha z_0}}{z_0^2} P \left| \int \int R(x, y, z_0)[s(x, y) ** e^{i(kr^2/2z_0)}]^2 dxdy \right| \tag{9.66}$$

where P is the peak value of $\tilde{p}(t)$.

The calculation of SNR using (9.64) and (9.66) is complicated somewhat by the constant K. This depends on a variety of factors, including the piezoelectric constants of the transducer. Equation (9.66) can be restructured in terms of experimentally measured values using a water tank. Assume that a specular reflector with unity or known reflectivity is placed close to the transducer face so as to be within the near field. This is an easily performed experiment using a material, such as a metal, whose acoustic impedance is very different than water. Since the reflector is in the near field, no diffraction effects are involved. In addition, the attenuation in the water tank, over the short path, is negligible. The peak reference signal E_r with $R = 1.0$ and no diffraction or attenuation is given by

$$E_r = KP \iint [s(x, y)]^2 dx dy. \tag{9.67}$$

Using this measured value, the value of E_0 is given by

$$E_0 = \frac{e^{-2\alpha z_0} E_r}{z_0^2 \iint [s(x, y)]^2 dx dy} \left| \iint R(x, y, z_0)[s(x, y) ** e^{i(kr^2/2z_0)}]^2 dx dy \right|. \tag{9.68}$$

This expression can be used for E_0 in (9.61) to find the signal-to-noise ratio based on experimentally measured values.

The integral expressions in (9.66) and (9.68) represent the product of the diffraction patterns of the source and the reflectivity at plane z_0. If the reflectivity function R, representing the object being studied at plane z_0, is small compared to the beam size, the integration is essentially over R itself. Conversely, if the reflectivity function, such as a lesion, is large compared to the beam pattern, the integration is effectively over the beam pattern and is independent of the size of the object.

SPECKLE NOISE

The noise studied thus far is that of electrical noise at the input of the system. Unfortunately, ultrasonic imaging has substantial coherence properties. This results in the introduction of a spatial noise component known as "speckle" [Burckhardt, 1978]. The origin of this component is seen if we model our reflectivity function as an array of scatterers. Because of the finite resolution, at any time we are receiving from a distribution of scatterers within the resolution element. These scattered signals add coherently; that is, they add constructively and destructively depending on the relative phases of each scattered waveform.

The noise properties of speckle are based on the statistical nature of the distribution of the sum of sinusoids reflected from the randomly distributed scatterers. The resultant phasors add in a random walk distribution. If there are a large number of scatterers within each resolution element, and the received phases are uniformly distributed from 0 to 2π radians, the envelope amplitude E obeys a Rayleigh probability density function given by

$$p(E) = \frac{2E}{\overline{E^2}} \exp\left(\frac{-E^2}{\overline{E^2}}\right) \tag{9.69}$$

where $\overline{E^2}$ is the average of the envelope squared. We define the signal-to-noise ratio as the ratio of the mean of the envelope to its standard deviation as given by

$$\text{SNR} = \frac{\bar{E}}{\sqrt{\overline{E^2} - \bar{E}^2}}. \tag{9.70}$$

We use a fundamental property of Rayleigh probability distributions which relates the mean of the envelope square to the square of the mean, as given by

$$\overline{E^2} = \frac{4}{\pi}\bar{E}^2. \tag{9.71}$$

Substituting this in (9.70) yields a signal-to-noise ratio of

$$\text{SNR} = \frac{\bar{E}}{\sigma_E} = \left(\frac{\pi}{4 - \pi}\right)^{1/2} = 1.91. \tag{9.72}$$

This relatively low ratio emphasizes the importance of this noise source. It often overrides the system electrical noise and is especially visible in larger organs consisting of uniformly distributed scatterers such as the liver. It is not yet well understood how much this noise component contributes to reducing the diagnostic accuracy of the image.

One mechanism of reducing this speckle noise is the summation of a number of images of the same object, each with independent speckle patterns. This will reduce the noise by the square root of the number of images. These independent images can be obtained by acquiring the data from different angular views. For example, compound scanning, illustrated in Fig. 9.13, reduces the speckle noise by acquiring and combining views of the same region from different angles.

PROBLEMS

9.1 A region of the body is modeled as a uniform array of scatterers. A fluid-filled nonreflecting cyst is contained within the volume so that the reflectivity is given by

$$R(x, y, z) = \text{comb}\left(\frac{x}{A}\right)\text{comb}\left(\frac{y}{A}\right)\text{comb}\left(\frac{z}{A}\right)\left(1 - \text{rect}\frac{x^2 + y^2 + (z - z_0)^2}{B}\right).$$

(a) Plot the estimated reflectivity versus depth in the A mode using an $L \times L$ transducer at the origin where $L < A \ll B$. Assume geometric imaging where diffraction is ignored. Assume that the attenuation has been compensated for and that the effective pulse envelope is rectangular with a period τ where $c\tau/2 < A$.

(b) Repeat part (a) where $3A < c\tau/2 < 4A$.

9.2 In modeling the reflectivity of the body we have assumed a weakly reflecting region where both the energy lost due to the reflection and multiple reflections can be ignored. Assume a simple one-dimensional planar model containing two reflecting surfaces which is modeled as

$$R(x, y, z) = R_1\delta(z - z_1) + R_2\delta(z - z_2).$$

(a) Find the estimated reflectivity $\hat{R}(0, 0, z)$ using a perfect delta function as the effective pulse envelope taking both the energy loss in reflection and multiple reflections into account. Ignore attenuation and calculate only the first received multiple reflection.

(b) Find the value of $\hat{R}$ using $R_1 = R_2 = 0.1$. Compare this to $\hat{R}$ using the weakly reflecting assumption.

9.3 An ultrasonic pulse with envelope $a(t) = \text{rect}\,(t/T)$ is transmitted through the structures shown in Fig. P9.3. Plot the envelope of the received signal, labeling amplitudes and times. Use the weakly reflecting assumption and neglect multiple reverberations.

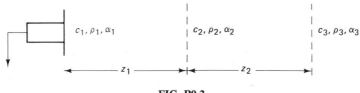

FIG. P9.3

9.4 In an ultrasonic imaging system, two isotropic point scatterers are in the $y = 0$ plane each a distance z from an $L \times L$ square transducer. We define two point images as being separable if the points are separated by at least the distance between the first zeros of the impulse response.

(a) What is the minimum separation of the points in the x direction to provide separable images where $z = z_1$ in the geometric near field and $z = z_2$ in the far field?

(b) What is the transducer size L that will achieve the same minimum separation of points in both the near and far fields at $z = z_1$ and z_2?

9.5 An ultrasonic imaging system has an additive noise power component $\overline{e_n^2}$. A diffusely reflecting disk having a reflectivity $A\text{circ}\,(r/r_0)\,\delta(z - z_0)$ is addressed with an on-axis circular transducer of radius r_t where $r_t > r_0$. The medium has a uniform attenuation α. Calculate the ratio of the signal-to-noise ratio in the near field at $z = z_1$ to that of the far field, where $z = z_2$. Assume geometric imaging for depth z_1 and Fraunhofer behavior for depth z_2.

10

Ultrasonic Imaging
Using Arrays

The various systems described in Chapter 9 all used a single transducer that was manually scanned to provide a two-dimensional image. These systems lack two desirable characteristics: real-time imaging and dynamic focus. Although real-time operation can be achieved by rapidly oscillating mechanical systems, it is presumed that an electronic scanning approach, with a stationary transducer array, is more desirable from the point of view of size, weight, and reliability. Dynamic focus, to overcome some of the basic diffraction problems illustrated in Figures 9.7, 9.8, and 9.9, can be achieved only through electronically controlled transducer arrays.

In our analysis we lean heavily on the results of Chapter 9 in developing the impulse response due to diffraction and the various considerations of attenuation, velocity, and reflectivity. We will consider all array configurations that are presently used or whose use is being considered [Macovski, 1979].

IMAGING ARRAYS

Imaging arrays are defined as transducer arrays which are in an image plane. This is in distinction to the arrays we will subsequently consider where the array is in a nonimaging plane and the various transducer signals are delayed and summed to provide image information.

A basic imaging array is shown in Fig. 10.1. The system shown uses near-field or collimated imaging. The individual transducers shown are fired in

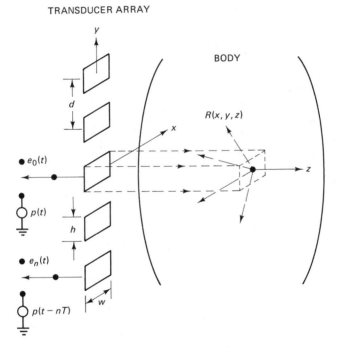

FIG. 10.1 Collimated imaging array.

sequence. Intially, we assume that we are operating within the near field of the individual transducers so that the propagated wave is a geometric extension of each transducer $s_n(x, y)$. Using rectangular transducers each individual source is described by

$$s_n(x, y) = \text{rect}\left(\frac{x}{w}\right) \text{rect}\left(\frac{y - nd}{h}\right). \tag{10.1}$$

As shown in Fig. 10.1, each transducer is driven by the pulse $p(t)$ in a sequence separated by time T. This time, as indicated in Chapter 9, must be greater than the maximum round-trip time $2z_{max}/c$, where z_{max} is the maximum depth. Each received pulse $e_n(t)$ is subjected to time-varying gain and envelope detected to provide a signal identical to that of equation (9.4). These signals are summed to provide an overall output given by

$$e_c(t) = K \sum_{n=-N/2}^{N/2} \left| \int \int \int R(x, y, z) s_n(x, y) \tilde{p}\left(t - nT - \frac{2z}{c}\right) dx\,dy\,dz \right| \quad (10.2)$$

where, as in Chapter 9, $e_c(t)$ is the gain-compensated, envelope-detected signal, $R(x, y, z)$ is the three-dimensional reflectivity of the object, and $\tilde{p}(t)$ is the received pulse as modified by the various linear parameters of the system, including propagation and filtering.

This array is an attempt to provide a cross-sectional image of the reflectivity in the $x = 0$ plane. When $e_c(t)$ is synchronously displayed, the resultant estimate of the reflectivity is given by

$$\hat{R}(0, y, z) = K \left| R(x, y, z) *** \tilde{p}\left(\frac{2z}{c}\right) \sum_{n=-N/2}^{N/2} s_n(x, y) \right| \quad \text{evaluated at } x = 0.$$

$$(10.3)$$

As before, the triple asterisk refers to a three-dimensional convolution. Each line in the reflectivity image is blurred by $s_n(x, y)$ in the lateral dimensions and $\tilde{p}(2z/c)$ in the depth dimension. Clearly a high-resolution image would require a relatively short pulse and a large number of relatively small, closely spaced transducers. However, as the transducers become smaller, our assumption of collimated imaging, which ignores diffraction, becomes less and less accurate.

In considering diffraction we will use the steady-state approximation of equation (9.21), which assumes that the pulse envelope arrives at each lateral position at the same time. We make use of the entire diffraction analysis of Chapter 9 and use the result of equation (9.34), the resultant signal using the Fresnel approximation. Applying this to the collimated array, the compensated detected signal is given by

$$e_c(t) = K \sum_{n=-N/2}^{N/2} \left| \int \int \int R(x, y, z) \left[s_n(x, y) ** \exp\left(i\frac{kr^2}{2z}\right) \right]^2 \right.$$
$$\left. \times \tilde{p}\left(t - nT - \frac{2z}{c}\right) dx\,dy\,dz \right| \quad (10.4)$$

where $r^2 = x^2 + y^2$. The lateral response has been degraded by diffraction as noted by convolution with the quadratic phase factor. The estimate of the reflectivity, including diffraction, is given by

$$\hat{R}(0, y, z) = K \left| R(x, y, z) *** \tilde{p}\left(\frac{2z}{c}\right) \sum_{n=-N/2}^{N/2} \left[s_n(x, y) ** \exp\left(i\frac{kr^2}{2z}\right) \right]^2 \right|$$

$$\text{evaluated at } x = 0. \quad (10.5)$$

This expression is valid for essentially all depth ranges, within the fairly broad limits of the Fresnel approximation.

A commercial imaging array is shown in Fig. 10.2 together with a typical image of a fetal head.

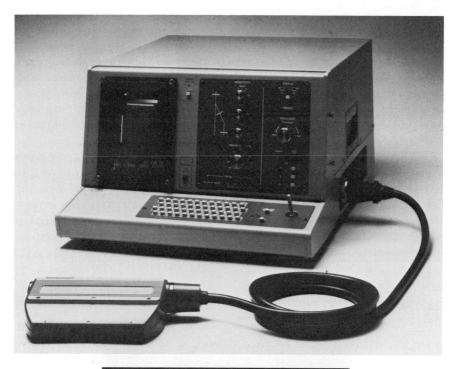

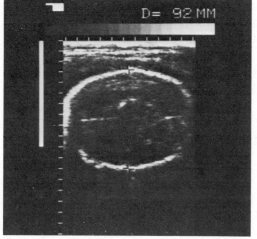

FIG. 10.2 Commercial imaging array and a typical image of a fetal head. (Courtesy of Siemens Gammasonics, Inc., Ultrasound Division.)

LIMITATION OF IMAGING ARRAYS

The imaging array of Fig. 10.1 has proven extremely useful in medical imaging applications involving superficial structures, including the eyes and major vessels close to the surface, such as the carotid arteries. In these regions relatively short wavelengths of about 0.2 mm can be used. For deep body imaging, the attenuation problem forces the use of longer wavelengths with the associated diffraction problems.

In addition to these diffraction problems, the field of view of imaging arrays is limited to the geometric extension of the array itself. In many medical imaging problems it is desired to study a region larger in extent than that of the array. This is particularly true of the heart, where a relatively small anatomical window, the cardiac notch, limits the size of the array to approximately 2.0 cm. A solution to both the lateral resolution problem and the field-of-view problem is electronic deflection and focusing, our next topic.

ELECTRONIC DEFLECTION AND FOCUSING

In electronic deflection and focusing systems, or *phased array systems* as they are often called, the transducer array is placed in a nonimaging plane. Unlike the imaging system, each transducer receives signals from every point in the field of view. The outputs of each transducer are appropriately delayed and summed so as to represent the energy reflected from a specific point. In this manner the signals from the desired point undergo constructive interference, while those from the other points undergo destructive interference. A variety of array configurations will be considered.

LINEAR ARRAY

A basic linear array imaging system is shown in Fig. 10.3. Using the controlled delay elements, the beam pattern is steered and focused so as to sequence through the region of interest [Somer, 1968]. Initially, we consider the use of the array solely in the receiving mode. We assume that the object is isotropically insonified by, for example, a small transducer at the origin of the array. We then calculate the selectivity of the receiver pattern.

For convenience in the analysis, we initially limit ourselves to the far-field region in the x dimension. In the y dimension, in linear array systems, there are no phased array imaging properties. In most cases the height h of the transducer

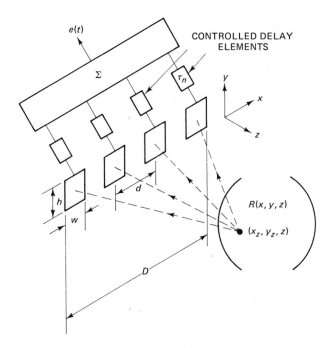

FIG. 10.3 Linear array imaging system, using controlled delays.

is made large enough such that the beam remains almost collimated in the y dimension throughout the depth of the volume of interest. In that case the linear array produces a sector-scan configuration with the beam deflected and focused in the x direction, and collimated in the y direction, as shown in Fig. 10.4.

To calculate the receiver response, we start with the general steady-state Fresnel expression for the impulse response at plane z, repeating (9.31), as given by

$$h(x_z, y_z) = h(x, y, z) = e^{ikz} \frac{s(x_0, y_0)}{z} ** \exp\left(i\frac{kr_0^2}{2z}\right).$$ (10.6)

Since we are considering the far field for the x components, we can approximate this pattern as

$$h(x_z, y_z) = \frac{e^{i\Omega}}{z} \mathcal{F}\{s_x(x_0)\}\left[s_y(y_0) * \exp\left(i\frac{ky_0^2}{2z}\right)\right]$$ (10.7)

where $\Omega = k(x_z^2/2z + z)$. It is assumed that our transducer source $s(x, y)$ is separable into the product $s_x(x)s_y(y)$. If we further assume that the height of the transducer is large enough to remain collimated throughout the depths of interest, the bracketed expression is simply replaced by $s_y(y)$. Using the rectangu-

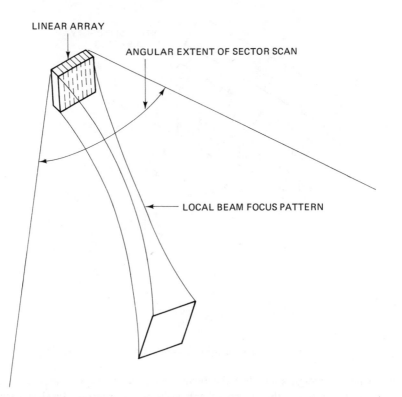

FIG. 10.4 Field pattern of a linear array sector scan, showing the result of deflection and focus to a specific region.

lar transducer array of Fig. 10.3, we have a source distribution given by

$$s(x, y) = \left\{\left[\operatorname{rect}\left(\frac{x}{D}\right)\operatorname{comb}\left(\frac{x}{d}\right)\right] * \operatorname{rect}\left(\frac{x}{w}\right)\right\}\operatorname{rect}\left(\frac{y}{h}\right). \qquad (10.8)$$

Convolution with the comb function replicates the individual transducer patterns, rect (x/w), at a spacing d. The overall width is defined by rect (x/D) with the height defined by rect (y/h). Applying (10.7) we obtain the field pattern as

$$h(x_z, y_z) = e^{i\Omega}\left\{\left[\frac{D}{z}\operatorname{sinc}\left(\frac{Dx_z}{\lambda z}\right) * d\operatorname{comb}\left(\frac{dx_z}{\lambda z}\right)\right]w\operatorname{sinc}\left(\frac{x_z w}{\lambda z}\right)\right\}$$

$$\times\left[\operatorname{rect}\left(\frac{y_0}{h}\right) * \exp\left(i\frac{ky_0^2}{2z}\right)\right]. \qquad (10.9)$$

Using the relationship

$$d\operatorname{comb}\left(\frac{dx_z}{\lambda z}\right) = \sum_{n=-\infty}^{\infty}\delta\left(\frac{x_z}{\lambda z} - \frac{n}{d}\right) \qquad (10.10)$$

we have

$$h(x_z, y_z) = e^{i\Omega\frac{Dw}{z}}\left\{\operatorname{sinc}\left(\frac{x_z w}{\lambda z}\right)\sum_{n=-\infty}^{\infty}\operatorname{sinc}\left[D\left(\frac{x_z}{\lambda z} - \frac{n}{d}\right)\right]\right\}$$

$$\times\left[\operatorname{rect}\left(\frac{y_0}{h}\right) * \exp\left(i\frac{ky_0^2}{2z}\right)\right]. \qquad (10.11)$$

The pattern in the y direction is the classical Fresnel pattern approximated in Fig. 9.8, where it is initially collimated and oscillatory and then diverges at an angle λ/h. The far-field pattern in the x direction from equation (10.1) is shown in Fig. 10.5. Of course, this represents the pattern without deflection

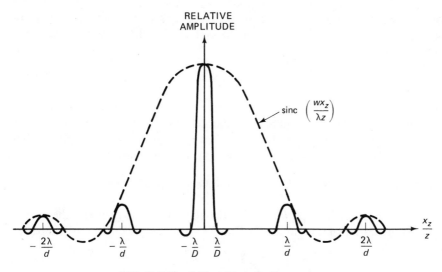

FIG. 10.5 Far-field pattern of a linear array.

since, as yet, no delays have been applied to the transducer outputs prior to summation. The x axis is conveniently normalized to x_z/z, which closely approximates the angle of the beam. As indicated in equation (10.11), the pattern in the x direction is an infinite series of sinc functions weighted by an overall envelope function sinc $(x_z w/\lambda z)$. For $w = d$, the array is a continuous transducer of extent D and thus reduces to the single response at the axis, known as the main lobe. In this case the zeros of sinc $(x_z w/\lambda z)$ occur at the undesired responses known as the "grating lobes" because of the similarity to the diffraction orders of an optical grating.

Each individual lobe, because of the sinc response, has an effective angular width or resolution of approximately λ/D, where the distance between zeros is $2\lambda/D$. The grating lobes occur at angular spacings of λ/d from the desired main lobe response. These result in undesired spurious responses since they receive energy from angles other than the direction of the main lobe. The full extent of this grating lobe problem is appreciated when we consider the deflection of the pattern.

LINEAR ARRAY WITH DEFLECTION

The controlled delays are used to steer or deflect the main lobe over an angular range to provide a sector scan. In general, the delay element τ_n at each transducer in Fig. 10.3 is equivalent to a convolution of each transducer signal with $\delta(t - \tau_n)$, where τ_n is a function of the array x coordinate. In the steady-state approximation to diffraction theory, this represents a phase shift of $\exp(i\omega_0\tau_n)$, where ω_0 is the center frequency of the system. For deflecting the beam, τ_n is made proportional to x as given by

$$\tau_n = \beta \frac{nd}{c} \tag{10.12}$$

where nd is the x coordinate of the center of each transducer and, as will be shown, β represents the resultant angular deflection of the beam pattern. Thus the phase shift at each transducer $\exp(i\omega_0\tau_n)$ is given by $\exp(ik\beta nd)$, where $k = \omega_0/c$. The field pattern in the x direction, $h(x_z)$, with these delays, is given by

$$h(x_z) = \frac{e^{i\Omega}}{z} \mathcal{F}\left\{ \left[\text{rect}\left(\frac{x_0}{D}\right) \text{comb}\left(\frac{x_0}{d}\right) \exp(ik\beta x_0) \right] * \text{rect}\left(\frac{x_0}{w}\right) \right\} \tag{10.13}$$

where the product of $\exp(ik\beta x)$ with the comb function provides the discrete phase shifts $\exp(ik\beta nd)$. Taking the Fourier transform yields the far-field pattern

$$h(x_z) = e^{i\Omega}\frac{Dw}{z} \text{sinc}\left(\frac{x_z w}{\lambda z}\right) \sum_{n=-\infty}^{\infty} \text{sinc}\left[D\left(\frac{x_z}{\lambda z} - \frac{n}{d} - \frac{\beta}{\lambda}\right)\right]. \tag{10.14}$$

This far-field impulse response of the deflected pattern in the x direction is shown in Fig. 10.6. Here the main lobe has been deflected to an angle of β. The grating lobes continue to be separated from the main lobe by multiples of λ/d. As the beam is deflected, the array of responses move within the overall envelope. As shown, the desired central lobe is reduced in amplitude and one of the first-order grating lobes comes up amplitude. Thus, in the presence of deflection, grating lobes become a more serious problem because of their increased relative amplitude.

One obvious method of reducing these grating lobes is to limit the angular scan β. This, of course, limits the field of view of the imaging system, making it unacceptable for many procedures. A more general approach is to reduce the spacing d, which, for a given array width D, corresponds to a larger number of transducers. This forces the grating lobe to be further down on the overall

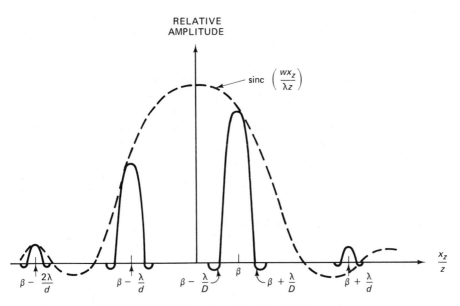

FIG. 10.6 Far-field pattern of a deflected linear array.

response until it becomes negligible. Another general approach is to use an insonification system, having its own spatial response, which minimizes the illumination of the regions of the grating lobes. Finally, as will be shown subsequently in this chapter, the use of short pulses, outside the quasi-steady-state approximation, reduces the relative peak amplitude of the grating lobes.

Using the system previously described, with an isotropic insonification source and a sector scanner, we again provide an estimate of the reflectivity in the $y = 0$ plane, the plane of the sector scan. In general, a series of M scan lines are generated, each with a different deflection angle β. β is incremented by an amount $\Delta\theta$ following the round-trip time of each line. The incremental angular scan $\Delta\theta$ should be less than the angular resolution λ/D to ensure that the system is adequately sampled. The resultant reflectivity estimate of the system is given by

$$\hat{R}(x, 0, z) = K \left| R(x, y, z)e^{i\Omega} *** \tilde{p}\left(\frac{2z}{c}\right) \left\{ \operatorname{sinc}\left(\frac{xw}{\lambda z}\right) \sum_{m=-M/2}^{M/2} \right. \right.$$

$$\left. \left. \times \sum_{n=-\infty}^{\infty} \operatorname{sinc}\left[D\left(\frac{x}{\lambda z} - \frac{n}{d} - \frac{m\Delta\theta}{\lambda}\right) \right] \left[\operatorname{rect}\left(\frac{y}{h}\right) * \exp\left(i\frac{ky^2}{2z}\right) \right] \right\} \right| \quad (10.15)$$

$$\text{evaluated at } y = 0.$$

As previously indicated, this is the system response for the case of isotropic insonification. Often the same linear transducer array, with the same delays, is used for insonification. In that case the system response of (10.15) is modified by simply squaring the lateral response terms in the braces. This then provides the required round-trip performance.

It must be emphasized that, in the x direction, we have assumed that we are in the far-field or Fraunhofer region. In this region the system performance cannot be improved through focusing. The far-field patterns represent the diffraction limit, or the best resolution attainable. In the near field, however, at distances less than D^2/λ, the transducer pattern is convolved with a quadratic phase factor. In this region, the response can be significantly improved through focusing to provide the desired diffraction limit.

LINEAR ARRAY WITH FOCUSING

In the Fresnel region the field pattern of the linear array in the x direction is given by

$$h(x_z) = \frac{e^{ikz}}{z} \left\{ \left[\text{rect}\left(\frac{x_0}{D}\right) \text{comb}\left(\frac{x_0}{d}\right) \exp\left[i\omega_0 \tau(x_0)\right] \right] * \text{rect}\left(\frac{x_0}{w}\right) \right\} * \exp\left(i\frac{kx_0^2}{2z}\right)$$

$$(10.16)$$

where $\tau(x_0)$ is the functional variation of the delays τ_n along the transducer x axis. In the case of deflection $\tau(x_0)$ was equal to $\beta x_0/c$, or $\beta nd/c$, resulting in a deflection angle β. For focusing, we need an additional delay structure to compensate for the quadratic phase factor [as is normally done by lenses, as in (9.46)]. We can restructure the Fresnel convolution of equation (10.16) into a Fourier transform form, as in equation (9.38), as given by

$$h(x_z) = \frac{e^{i\Omega}}{z} \mathcal{F} \left\{ \left[\left(\text{rect}\left(\frac{x_0}{d}\right) \text{comb}\left(\frac{x_0}{x}\right) \exp\left[i\omega_0 \tau(x_0)\right] \right) * \text{rect}\left(\frac{x_0}{w}\right) \right] \exp\left(i\frac{kx_0^2}{2z}\right) \right\}$$

$$(10.17)$$

which can be modified as

$$h(x_z) = \frac{e^{i\Omega}}{z} \mathcal{F} \left\{ \sum_{n=-N/2}^{N/2} d \, \text{rect}\left(\frac{x_0 - nd}{w}\right) \exp\left[i\left(\frac{k}{2z}x_0^2 + kc\tau(nd)\right)\right] \right\} \quad (10.18)$$

where the overall width of the array rect (x/D) has been replaced by a finite sum N of array elements. In general, the function of τ_n is to provide deflection, where $\tau_n = \beta nd/c$, and to provide focusing where it is used to cancel the quadratic phase factor in (10.18) and provide a diffraction-limited response. Thus the delay at each transducer $\tau(nd)$ is given by

$$\tau(nd) = \frac{\beta nd}{c} - \frac{\psi(nd)^2}{c}. \qquad (10.19)$$

For convenience we can ignore the deflection angle β and study the requirements of the quadratic phase shift ψ. By setting $\psi = 1/2z$ at each depth, we cancel the quadratic phase factor at the center of each transducer element, where $x = nd$. To achieve our desired focusing at each depth, that is, to provide diffraction-limited imaging at all depths, the quadratic phase factor must be eliminated so that the response in the x direction is the desired Fourier transform of the array aperture. At the edges of each transducer, where $x_0 = nd + w/2$, the phase factors reach their local maximum. The worst case occurs at the edge of the array where $x_0 = D/2$ and $(nd)_{\max} = D/2 - w/2$. If we use the sufficient condition that the resultant quadratic phase factor in (10.18) be small compared to 1 radian, we have the inequality

$$\frac{\lambda}{w} \gg \frac{D}{z}. \qquad (10.20)$$

This requirement is necessary in any case if we are to achieve a reasonable angular field of view. The angular field of the sector-scan pattern, as shown in Figs. 10.5 and 10.6, is determined by λ/w, the angular field of each transducer. The angle D/z represents an angular field whose largest extent is the size of the array itself. Thus (10.20) must be satisfied if the field of view is to encompass a region significantly larger than the array size.

We therefore effectively cancel the quadratic phase factor and provide focusing by having $\psi = 1/2z$. The focusing delays must vary dynamically with depth as the pulse propagates. Various time-varying delay elements, such as charge-coupled devices or switched delay lines, are used for this function. In general, the deflection delays are switched to a new value of β following each scan line, while the focusing delays are varied dynamically during the line scan. The delays specified for deflection and focusing in (10.19) can be negative. The actual physical delays $\tau(nd)$ must therefore have an added term T_0 to ensure that they are positive. This fixed delay at each element does not affect the responses and merely adds an overall delay to the received signal.

The use of dynamic focusing provides the response of equation (10.14) at all depths, not just in the far field. When the quadratic phase factor of equation (10.18) is effectively controlled, we again have the Fourier transform of the array of rectangular elements, resulting in the response of equation (10.14). With dynamic focusing, the array width D can be increased to provide improved lateral resolution without interfering with the far-field approximation. Unfortunately, there are often anatomical limits on the size of the array, such as the cardiac notch. Despite this, linear array sector-scan imaging of the heart using both electronic deflection and focusing has become a viable diagnostic product. A commercial sector-scan instrument together with a typical heart image are shown in Fig. 10.7.

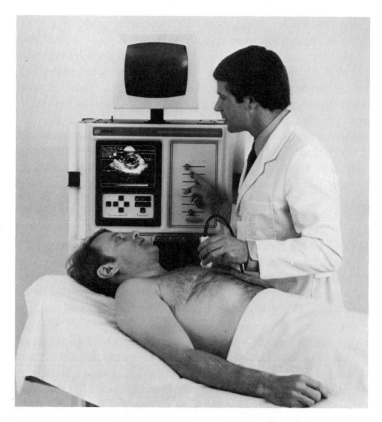

FIG. 10.7 Commercial sector-scan instrument, showing a typical heart scan. (Courtesy of Varian Associates.)

WIDEBAND RESPONSES OF A LINEAR ARRAY

The analysis of the linear array was made using the quasi-steady-state approximation to diffraction theory. For considerations of the wideband case, using relatively narrow pulses, we make use of the development in Chapter 9 [Norton, 1976]. For example, equation (9.51) represents the one-way response in the x direction using a rectangular transducer and a rectangular pulse envelope. The equations following (9.51) are an expansion of this result. This formulation is identical to the one for the main lobe ($n = 0$) of a linear array system without deflection, where $\beta = 0$. In equation (10.14) for the response of a linear array, setting $\beta = 0$ and $n = 0$, we have the product of a broad sinc pattern sinc $(x_z w/\lambda z)$ and a narrow sinc pattern sinc $(x_z D/\lambda z)$. Since $D \gg w$, we can neglect the envelope of the broad sinc pattern. This reduces the response to the Fourier transform of rect (x/D), making it identical to the wideband case studied in Chapter 9.

Thus the behavior of the main lobe, pointing straight ahead, is identical to that illustrated in Figures 9.10 and 9.11 for the single-transducer case. We again have the diagonal arms in the response, which are neglected in the quasi-steady-state treatment. The accurate wideband treatment also has a profound effect on the grating lobes. In evaluating the main-lobe response we set $n = 0$ and essentially assumed negligible overlap between the various orders. We follow the same practice in evaluating the amplitudes of the nth order. Using the same development as before, where $\tilde{p}(t) = $ rect (t/τ), we have the amplitude response due to the nth order as

$$u_n(x_z, t) = \frac{\exp(-i\omega_0 t)}{z} \mathcal{F} \left\{ \text{rect} \left(\frac{x}{D} \right) \exp \left(\frac{2\pi i n x}{d} \right) \text{rect} \left(\frac{x\phi - t'}{\tau'} \right) \right\} \quad (10.21)$$

where, as before, $\phi = x_z/z$, $\tau' = \tau c$, and $t' = (T_1 - t)c$. Here the complex exponential gives rise to a grating lobe centered at $\phi = n\lambda/d$.

To study the relative amplitudes of the grating lobes in response to short pulses, we set $t' = 0$, corresponding to the center of the temporal response in both cases. The ratio of the amplitude of the grating lobe to that of the main lobe is defined as R_n and is given by

$$R_n = \frac{u_n(n\lambda/d, 0)}{u_0(0, 0)}. \quad (10.22)$$

Using (10.21), this ratio is calculated as

$$R_n = \begin{cases} \dfrac{\tau' d}{n\lambda D} & \text{for } \tau' < \dfrac{n\lambda D}{d} \\[3mm] 1 & \text{for } \tau' \geq \dfrac{n\lambda D}{d}. \end{cases} \quad (10.23)$$

This relationship can be given additional physical meaning by noting that the number of elements in the array is

$$N = \frac{D}{d} \quad (10.24)$$

and the number of cycles in the pulse is

$$m = \frac{\tau'}{\lambda}. \quad (10.25)$$

Substituting in (10.23), we have

$$R_n = \begin{cases} \dfrac{m}{Nn} & \text{for } \dfrac{m}{Nn} < 1 \\[3mm] 1 & \text{for } \dfrac{m}{Nn} \geq 1. \end{cases} \quad (10.26)$$

Thus for the important first grating lobe, $n = 1$, the amplitude is reduced by the ratio of the number of cycles in the pulse to the number of elements in the array. Typically, this ratio is about an order of magnitude.

This reduction in the peak amplitude of the grating lobes is brought about by a "smearing out" of the grating lobe response. This is illustrated in Fig. 10.8. In addition to the grating lobe reduction, Fig. 10.8 illustrates improved "smoothness" of the main-lobe response brought about by a reduction in the sidelobes of the sinc pattern. In narrowband systems with relatively long pulses having patterns similar to that of Fig. 10.5, a reduction in these sidelobes can be achieved by apodization techniques. Here the overall weighting function of the transducer array is modified from a rect function to a smoother function whose resultant Fourier transform is relatively free of the sidelobes.

In the wideband analysis, as indicated earlier, we have neglected the overall envelope pattern due to the individual transducer responses. This approximation is valid if these transducer widths w are relatively narrow. Otherwise, this represents an additional factor in the grating-lobe amplitude, which varies with the deflection angle β.

TWO-DIMENSIONAL ARRAY SYSTEMS

The system described thus far involves processing in one dimension only. The beam pattern in the y direction was dictated by the classic diffraction patterns described in Chapter 9. To provide diffraction-limited performance in both lateral dimensions, a two-dimensional array must be used.

RECTANGULAR ARRAY

A two-dimensional rectangular array can be used to provide a sector scan with dynamic focus in both lateral dimensions. For example, the beam, as before, can be deflected in the x direction by a controlled amount β. At each beam position, dynamic focus can be applied in both lateral dimensions by again applying a quadratic time delay to cancel the quadratic phase factor. Simply extending the previous results, we apply a time delay τ_{mn} to each element:

$$\tau_{mn} = T_0 + \frac{1}{c}[\beta nd - \psi(nd)^2 - \gamma(md)^2] \tag{10.27}$$

where $-\gamma(md)^2$ is the additional delay given each element in the y direction to provide focusing. Both ψ and γ are varied proportional to $1/z$ as the wavefront propagates to provide dynamic focusing at each depth plane. As can be seen from (10.27), each of the $N \times N$ elements requires an independent delay control to achieve the focusing in both dimensions. At the present state of the art this

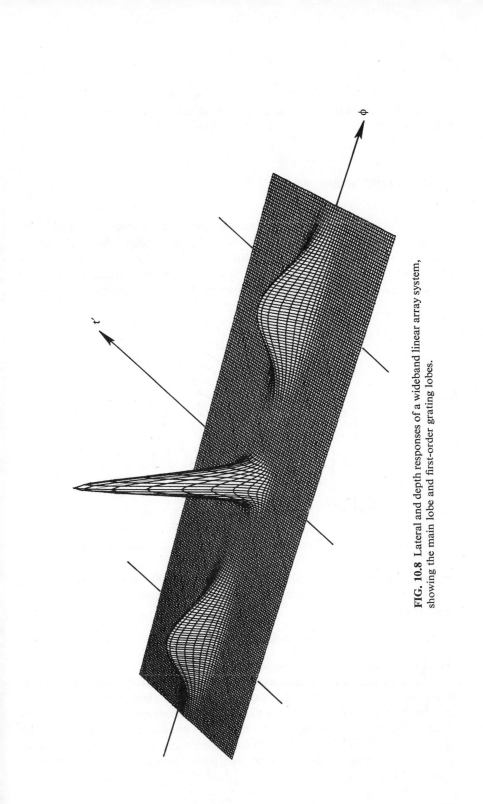

FIG. 10.8 Lateral and depth responses of a wideband linear array system, showing the main lobe and first-order grating lobes.

φ

t′

represents excessive cost and complexity. Other, less complex approaches have been used. For example, a weakly focused lens can be affixed to a linear transducer array to provide improved lateral resolution for a range of depths.

CONCENTRIC RING ARRAY

The complexity of the controlled delay requirements is greatly reduced if the delays are used solely to provide focusing. If the beam is not electronically deflected, delays are required which are proportional to r^2, the square of the radial distance to the transducer axis. Systems of this type are available which use an array of concentric rings as shown in Fig. 10.9. The controlled delays

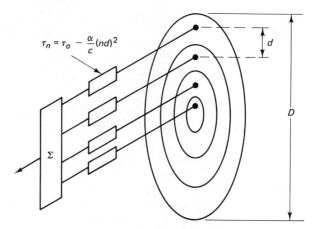

$$\tau_n = \tau_o - \frac{\alpha}{c}(nd)^2$$

FIG. 10.9 Dynamic focusing system, using concentric transducer rings.

are time varied so as to again cancel the quadratic phase factor in equations (9.31) and (9.38). The resultant lateral response, making use of radial symmetry, is given by

$$h(r_z) = \frac{e^{iv}}{z}\mathfrak{F}\left\{\sum_{n=1}^{N} s_n(r_0) \exp\left[-ik\alpha(nd)^2\right]\exp\left[i\left(\frac{k}{2z}\right)r_0^2\right]\right\} \qquad (10.28)$$

where $s_n(r_0)$ represents each annular ring.

As with the linear array, we dynamically set $\alpha = 1/2z$ at each depth, providing an exact cancellation of the quadratic phase shift at the center of each annular ring, where $r_0 = nd$. Using the same approximations as those previously described for the linear array, under the same conditions as in (10.20), we can assume that the quadratic phase shift is adequately canceled over the entire array. Under these conditions, with contiguous annular rings, the lateral or radial impulse response (10.28) becomes

$$h(r_z) = \frac{e^{iv}}{z}\mathfrak{F}\left\{\text{circ}\left(\frac{r_0}{D/2}\right)\right\} = e^{iv}\frac{\pi D^2}{2z}\frac{J_1(kDr_z/2z)}{(kDr_z/2z)} \tag{10.29}$$

where the circ function, as previously described, is unity for $0 \leq r_0 \leq 1$ and zero otherwise, and J_1 is the Bessel function of the first kind and first order. The resultant pattern is often referred to as a "jinc" function because of its similarity to the sinc function, as illustrated in Fig. (2.3). Its effective resolution, or width of the main lobe, is again approximately $\lambda z/D$, as with the linear array.

The concentric ring array, as shown, provides dynamic focusing but not deflection. The deflection of the beam is provided mechanically either by translating or tilting the ring array. The former provides a rectangular sectional image format, while the latter provides a sector scan.

ANNULAR RING ARRAY TRANSMITTER

The concentric ring array fails to provide both electronic scanning and dynamic focusing. To accomplish this without the complexity of a rectangular array, we make use of the combination of a separate transmit and receive array pattern with the resultant pattern being the product. It must be emphasized, however, that the transmitter pattern is not subject to dynamic control. Unlike the receiver array, once the pulsed waveform is launched we lose control of the pattern, so that it cannot be varied with depth. To overcome this, we make use of the unique properties of an annular ring transmitter pattern [Macovski and Norton, 1975]. An annular ring of radius R with infinitesimal thickness has a lateral impulse response given by

$$h(r_z) = \frac{e^{iv}}{z}\mathfrak{F}\left\{\delta(r_0 - R)\exp\left[i\left(\frac{k}{2z}\right)r_0^2\right]\right\}. \tag{10.30}$$

Applying the sifting property of the delta function, the quadratic phase factor comes out of the Fourier transform integral to form a constant phase factor, as studied previously. We are left with the Fourier transform of an annulus which is simply the kernel of the Hankel or Fourier–Bessel transform of (2.35). The resultant pattern is given by

$$h(r_z) = e^{iv}\frac{2\pi R}{z}J_0\left(\frac{2\pi Rr_z}{\lambda z}\right). \tag{10.31}$$

Thus the quadratic phase factor, in an annular ring configuration, is automatically removed. The resultant response of (10.31) is therefore the same in both the near and far field. This achieves an effective focusing which is depth independent.

To provide deflection, we again require a linearly varying phase shift exp $(ik\beta x_0)$. To provide this with an annular ring array, the array must be segmented so that individual delays $\tau_i = \beta x_i/c$ can be applied. Adding this linear phase factor in (10.30) provides a deflected pattern given by

$$h(x_z, y_z) = \frac{e^{iv}}{z} \mathcal{F} \left\{ \delta(r_0 - R) \exp\left[i\left(\frac{k}{2z}\right) r_0^2 \right] \exp(ik\beta x_0) \right\} \qquad (10.32)$$

$$= e^{iv} \frac{2\pi R}{z} J_0 \left(\frac{2\pi R \sqrt{(x_z - \beta z)^2 + y_z^2}}{\lambda z} \right) \qquad (10.33)$$

where we have used the shift relationship

$$\mathcal{F}\{\exp(ik\beta x_0)\} = \delta(x_z - \beta z). \qquad (10.34)$$

Equation (10.33) essentially assumed continuous segmentation of the annulus. With finite segmentation the response is modified similarly to that of the linear array, depending on the nature of the segmentation. As indicated, the response is simply translated an amount βz or deflected by an angle β.

The segmented annular ring provides a $J_0(\cdot)$ response at all depths which can be relatively simply deflected. This response has the desired narrow central lobe, but has a severe problem of sidelobes. It is a highly oscillatory response whose first sidelobe amplitude is 40% of the peak response and whose fourth sidelobe has a relative amplitude of 22%. A pattern of this type causes many false responses. Thus the uniform annulus would function poorly where the transmitter pattern is relied on to provide the required resolution in one of the lateral dimensions.

THETA ARRAY

This response can be modified by taking advantage of the fact that the desired transmitter pattern need only provide improved resolution in one dimension. The dynamically focused linear array can provide the desired pattern in the x direction with a relatively simple structure. The annular array can be modified, using angular weightings, to provide an improved response in the y direction. For example, using a $\cos^2 \theta$ weighting, we have

$$h(r_z, \phi) = \frac{e^{iv}}{z} \mathcal{F}\{\delta(r_0 - R) \cos^2 \theta\}. \qquad (10.35)$$

Using the polar transform relationship of equation (2.34), we have

$$h(r_z, \phi) = \frac{e^{iv}}{z} \left[J_0\left(\frac{kRr_z}{z}\right) - J_2\left(\frac{kRr_z}{z}\right) \cos 2\phi \right]. \qquad (10.36)$$

Using the Bessel function recursion identity

$$J_{n-1}(x) + J_{n+1}(x) = \frac{2nJ_n(x)}{x} \qquad (10.37)$$

we can rewrite (10.36) as

$$h(r_z, \phi) = e^{iv}\left[\frac{J_1(kRr_z/z)}{kRr_z} - \frac{1}{z}J_2\left(\frac{kRr_z}{z}\right)\cos^2\phi\right]. \qquad (10.38)$$

In this form we can appreciate that the response in the y direction ($\phi = 90°$, 270°) is the desired jinc function, which characterizes a full aperture system focused at the desired depth. The response in the x direction has significant sidelobe content. However, in the x direction, the overall pattern is mostly dominated by a dynamically focused linear array, as shown in Fig. 10.10.

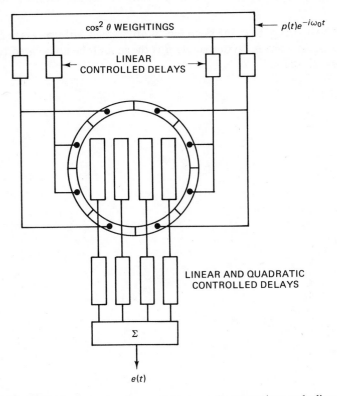

FIG. 10.10 Theta array, using a weighted annular transmitter and a linear array receiver.

This system, referred to as the *theta array*, achieves diffraction-limited resolution in both dimensions using a relatively simple structure. As shown, a set of linearly varying delays in the x direction is used to deflect the annular array to the same angle as that of the linear array. The linear array has both linear and quadratic delays for deflection and focusing.

PROBLEMS

10.1 Equation (10.11) is the on-axis impulse response of a linear array of rectangular transducers. Find the impulse response, $h(x_z, y_z)$, for a linear array of circular transducers of radii R and center-to-center spacing d.

10.2 Estimate the largest ratio of grating lobe amplitude to central lobe amplitude for a linear array system that has been deflected by 0.5 radian. The array consists of 20 contiguous elements of width $w = 1.0$ mm. Assume the steady-state approximation, operating in the far field, where $\lambda = 0.5$ mm.

10.3 As shown in Fig. 10.6, the main lobe is attenuated with increased deflection angle β. At what deflection angle β is the main lobe attenuated to 50% of its on-axis value?

11

Selected Topics
in Medical Imaging

In this chapter we consider a variety of medical imaging techniques which are not broadly involved in current clinical practice. Some of these are at the level of basic research, while others have advanced to the stage of initial clinical trials.

NUCLEAR MAGNETIC RESONANCE

Nuclear magnetic resonance, or NMR as it is often abbreviated, has recently been adapted to medical imaging. Some of the results, especially in providing cross-sectional images of the head, have been very promising, so that this modality is clearly worth considering. NMR requires subjecting the body to relatively intense magnetic fields. Thus far, these appear to be without any toxic effects, so that it has the advantage, over x-rays, of being free of ionizing radiation.

The basic NMR phenomenon [Bloch, 1946] has been used as an analytic tool in chemistry and physics since its discovery. The phenomenon is based on the magnetic moment present in a wide variety of organic and inorganic materials. To possess a magnetic moment the nuclei must contain an odd number of protons or neutrons. This requirment is met by a very large percentage of stable nuclei and radiosotopes. Most important for medical imaging, hydrogen possesses a magnetic moment and is by far the most active source of NMR signals among the elements.

The presence of a magnetic moment is equivalent to the nuclei being arrays of small magnets. When these are placed in an external magnetic field the magnetic moment tends to align itself parallel to the field. Since the nucleus is spinning, the magnetic moment responds to the external field like a gyroscope precessing around the direction of the field. The rotating or precessing frequency of the spins ω_0 is known as the *Larmor frequency* and is given by

$$\omega_0 = \gamma H \tag{11.1}$$

where γ is the gyromagnetic ratio, a property of the material, and H is the external magnetic field.

Equation (11.1) represents the fundamental relationship between the magnetic field and frequency for a given material. It is this relationship that forms the basis of various imaging modalities. Complex magnetic field distributions are used so that each spatial region has a unique magnetic field and thus a unique frequency. In the area of chemical analysis, in nonimaging systems, equation (11.1) is used for material analysis where a fixed magnetic field is applied to a small volume of interest. Each material in the sample represents a different frequency.

To perform either imaging or material analysis a signal at the precession frequency must be emitted by the material. This is accomplished by exciting the precession with a radio frequency rotating field in the x, y plane in addition to the static field in the z direction. The total vector field $\vec{H}$ is then given by

$$\vec{H} = H_0\hat{z} + H_1(\hat{x}\cos\omega_0 t + \hat{y}\sin\omega_0 t) \tag{11.2}$$

where $\hat{x}, \hat{y}$, and $\hat{z}$ are the unit vectors. The resultant precessing moment is shown in Fig. 11.1. The precession or tipping angle θ is given by

$$\theta = \gamma H_1 t_p \tag{11.3}$$

where t_p is the duration of the radio-frequency rotating field excitation.

When the excitation ceases, the rotating magnetic moment undergoes "free induction decay" as it decays to its equilibrium state. In this decay process a signal is emitted at the resonant frequency $\omega_0 = \gamma H$. It is this signal that is used in image formation. The signal is normally detected using the same coils that produced the rotating magnetic field H_1.

The resulting signal V is proportional to the hydrogen density of the material, since each spinning nucleus is contributing. Following the excitation, the magnetic moment returns to its equilibrium value with a time constant T_1, known as the longitudinal or spin-lattice relaxation time. In most imaging systems, as we will consider, repeated excitations are required of the same region.

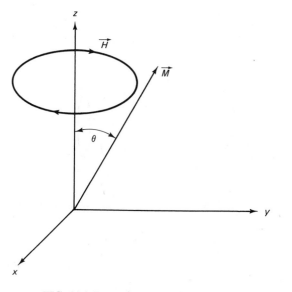

FIG. 11.1 Precessing magnetic moment.

When a region is reexcited that has not fully relaxed to its equilibrium value, the resultant free induction decay or FID signal amplitude V is diminished as given by

$$V = k\rho(1 - e^{t_a/T_1}) \qquad (11.4)$$

where k is a proportionality constant, ρ is the density of the material being imaged, usually hydrogen, and t_a is the time interval between excitations.

As can be seen, the resulting performance is a compromise between signal strength and imaging time. The longitudinal relaxation time T_1 is approximately 1 sec. Therefore, t_a must be a reasonable fraction of a second to provide a large fraction of the maximum signal $k\rho$. As a result, many imaging procedures require a few minutes of data acquisition time.

Note that for relatively short values of t_a, the resultant signal will depend both on ρ and T_1. These, however, are both important clinical properties of the material being studied. Measurements at two values of t_a can separate the values of ρ and T_1.

A variety of volumetric imaging methods can be used. One straightforward method, called *zeugmatography* [Lauterbur and Lai, 1980], involves acquiring an array of planar integrals of the volume at all angles and then reconstructing the activity of each voxel in the volume. The isolation of signals from particular planes is accomplished by adding a gradient field in different directions to the static field H_0, which is in the z direction. A representative geometry is shown in Fig. 11.2.

In addition to the static H_0 field and the radio-frequency H_1 field at ω_0 generated using coils, a relatively small gradient field H_g is added in different

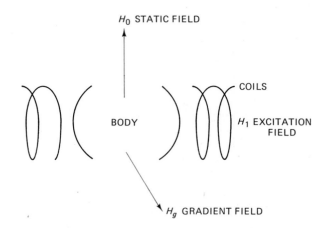

FIG. 11.2 Basic NMR imaging configuration.

directions. For illustrative purposes assume a gradient in the x direction given by

$$H_g = Gx \qquad (11.5)$$

where $H_g(x)$ is the added field at each x position. Using equation (11.1) we see that each yz plane, at different values of x, has its own nuclear resonant frequency, as given by

$$\omega_0(x) = \gamma(H_0 + Gx). \qquad (11.6)$$

An excitation signal H_1 can be used at a specific frequency $\gamma(H_0 + Gx_1)$ which will excite a specific yz plane represented by the specific x_1 value represented by that frequency. Thus, by a sequence of excitations at different frequencies, the integrated density of each plane can be derived. A more efficient approach utilizes a broadband excitation for H_1 where the excitation simultaneously provides the required 90° tipping angle for each frequency or plane. For example, an excitation of the type sinc $(t/\tau) \cos \omega_0 t$ will have a rectangular spectrum having a bandwidth $1/\tau$ centered at ω_0. Thus each yz plane, at each x value, produces the "free induction decay" signal at its own frequency. If we take a Fourier transform of the received signal, decomposing it into its frequency components, we will have an array of measurements representing the planar integration of the hydrogen density at each value of x. Thus a single wideband pulse results in the acquisition of a set of planar integrals.

This process can be repeated with the gradient shifted to all possible directions to provide a complete set of planar integrals. These can then be used to reconstruct the hydrogen density of every voxel in the volume using the techniques of reconstruction from projections.

One approach to the reconstruction [Lauterbur and Lai, 1980] is to first derive an array of two-dimensional projections of the volume. For example, assume that a series of measurements are taken with the gradient direction normal to the x axis. Thus the gradient direction is incrementally rotated around the x axis. In each case, an array of planar integrals are formed which are parallel to the x axis. These represent line integrals of the projections of the volume in the x direction. Using the classic reconstruction technique described in Chapter 7, the projection of the volume in the x direction can be calculated. This process can be repeated, each time developing a two-dimensional projection of the volume in an array of planes parallel to the y axis.

Once this set of projections exists, in a cylindrical geometry about the y axis, we have the required information to reconstruct any planar cross section parallel to the xz plane. Using the calculated projection data at all angles, we again use the technique of reconstruction from projections to reconstruct any element in the plane. Thus the information is processed using two successive applications of reconstruction: first to obtain the projections from the planar integrals, and then to reconstruct the individual planes from the projection calculation.

A second generic approach to NMR is the multiple sensitive point method [Andrew, 1980]. This method utilizes an alternating gradient rather than the previously described static gradient. The resultant NMR signal is then modulated by the frequency of alteration, typically of the order of 100 Hz. If this frequency is filtered out of the resultant received signal, the average value represents the planar integral of the density at the narrow slice having a zero alternating field. We have thus limited the acquisition to a single plane. Simultaneously, an alternating gradient field of a different frequency can be applied normal to the first gradient. After filtering both frequencies, the resultant signal represents the density of single line, corresponding to the intersection of the two zero alternating field planes.

Many variations are possible on this general theme for reconstructing each voxel. The acquired sensitive line can be moved through a plane at all angles and positions with the data used to reconstruct each pixel in the plane. Alternatively, a third alternating field at the third perpendicular axis at another frequency can be used so as to cause the filtered received signal to represent a single point. By manipulating the fields this single point can be scanned throughout the volume. A more efficient, less time-consuming approach is to use a static field in this third axis. In this case, each point in the line will produce a different frequency when excited with a wideband pulse. Again, a Fourier transform can be used to provide the response at each frequency, simultaneously providing the density of each point on the line.

Another general imaging method we will consider makes use of a fundamental property of NMR as given by equation (11.3). As indicated, the precession angle θ is determined by the strength of the rotating field H_1 and the time duration of the H_1 pulse. This method is called the selective iradiation process [Crooks, 1980]. Thus far we have considered excitations involving $\theta = 90°$, resulting in a free induction decay (FID) signal. This approach involves $\theta = 180°$, known as an *inverting excitation*. This excitation does not produce a free induction decay signal. Instead, the resulting magnetic moment returns to equilibrium with the relaxation time T_1. Therefore, if a 90° excitation is applied at a time t_b after the 180° inverting excitation, the resultant FID signal amplitude is given by

$$V = k\rho(1 - 2e^{-t_b/T_1}).$$ (11.7)

Note the factor of 2 resulting from the initial inversion excitation.

At first a planar integration is acquired as previously described by using a static gradient and a 90° excitation pulse at the appropriate frequency. The resultant signal representing the integrated density in a plane is stored. After the system reaches equilibrium we apply a second static gradient and excitation pulse in quadrature with the first plane. This second excitation pulse, however, is doubled in amplitude and/or time so that $\theta = 180°$, thus inverting the magnetic moments. This excitation produces no received free induction decay signal. The first excitation is then repeated after a time t_b to again obtain the integration of the same plane as modified by equation (11.7).

The integrated values of the plane taken before and after the inverting excitation are subtracted. All the values in the plane will cancel except for the line of intersection of the plane with the quadrature plane having the inverting excitation. In this line, the preconditioned signal will be different, and therefore not cancel. If the time between the inversion and the readout is sufficiently short, the signal along the line of intersection will be the negative of the previous signal, so that the subtraction will double the amplitude of the selected line. In any case, the output will be limited to the line of intersection. This line can again be scanned throughout any plane of interest to provide an appropriate reconstruction.

The 180° inversion excitation can be used following a 90° excitation to provide another imaging approach. The free induction decay signal resulting from a 90° excitation has a relatively short time constant because of the local inhomogeneity of the fields caused by the gradients. In general, each nuclei is precessing at a slightly different frequency. The resultant phase shifts cause destructive interference, resulting in the rapid decay. However, the 180° inversion excitation following the 90° excitation by a time interval t_c causes the various phase relationships between the individual nuclei to be reversed. A given spin which is lagging another in phase by an angle ψ is altered so as to lead in phase by the same angle. Therefore, following this 180° excitation the various precessing components that were out of phase return to being in phase after the time t_c. The resultant signal produced is called a *spin echo*. A simple analogy is a number of racing cars leaving a starting point at different speeds and slowly becoming "out of phase" after a time T. If their directions are reversed, after another time T, they will again be "in phase" at the starting line.

The spin echo signal has an amplitude somewhat less than the original FID. This loss is due to the spin-spin relaxation process whereby the interaction of each precessing nuclei with its neighbor causes some dephasing. This is a random process, unlike the dephasing due to the nonuniform field, and cannot be reversed. It is characterized by the time constant T_2, another important property of the material which could prove important in diagnosis. The resultant spin echo signal amplitude is given by

$$V = k\rho e^{-2t_c/T_2}. \tag{11.8}$$

This process can be used to measure the T_2 property of the material or as an alternative imaging system.

The spin echo phenomenon can be used in imaging by again isolating a line as the intersection between two planes. The first 90° excitation is applied to one plane and then, after a time t_c, the 180° inversion excitation is applied to the quadrature plane. Following another time interval t_c, a spin echo will be produced which represents only the line intersection of the two planes.

A novel approach to NMR imaging which appears quite promising is known as the *spin warp* system. It has the interesting distinction of using electrical and spatial Fourier transforms in the reconstruction process. Prior to acquiring data from a plane, or a set of planes, a gradient field is applied normal to the plane. This gradient momentarily causes frequency changes along the gradient. When the gradient is removed, and the signal is received, regions along the gradient will provide signals at different phases.

These phase variations will be cyclical with a periodicity based on the strength of the gradient. Effectively, the information in the plane is being decomposed into a specific spatial frequency based on the strength of the gradient. Therefore, a series of acquisitions at different gradient amplitudes results in a decomposition of the plane into its spatial frequency components normal to the gradient.

To define each pixel in the plane completely, information is required normal to the direction of the spatial frequency decomposition. This is accomplished using an additional gradient normal to the one described previously which is applied prior to data acquisition. This additional gradient is present during the acqusition time so that each line in the plane corresponds to a different emitted frequency. A Fourier transform of the temporal signal decomposes the plane in one direction while a spatial Fourier transform of the successive cyclical phase variations decomposes the plane in the orthogonal direction, defining each pixel.

We have discussed many approaches to imaging the density of the magnetic moments which effectively represent hydrogen density when the appropriate frequencies are used. As indicated, hydrogen provided by far the most intense signal, which allows imaging in a reasonable time. Other elements have magnetic moments. However, the reduced abundance and magnetic moment necessitates very long data acquisition times, so that images of other elements have yet to be obtained in a clinical setting.

In addition to measuring hydrogen density, the various decay-time constants can also be measured and used to form useful clinical images. Some examples of clinical images using NMR are shown in Fig. 11.3.

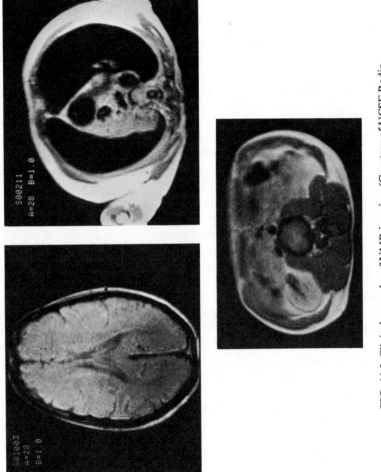

FIG. 11.3 Clinical examples of NMR imaging. (Courtesy of UCSF Radiologic Imaging Laboratory.)

DIGITAL SUBTRACTION RADIOGRAPHY

The principal motivation of this technique is the noninvasive study of vessels [Ovitt et al., 1978]. With invasive procedures where catheters are inserted into vessels, large amounts of iodinated contrast material are present, resulting in good vessel visualization, despite the intervening anatomical structures. In noninvasive studies, however, the very low iodine concentration is insufficient to be visualized in a normal radiograph. As a result, subtraction techniques are employed to eliminate the intervening tissue so that the iodine visualization is limited solely by noise. The subtraction technique involves obtaining data both before and after the administration of the contrast agent, and subtracting the result.

Let $\mu_t(x, y, z)$ be the attenuation coefficient of the tissue in the anatomical region under study. Let $\mu_c(x, y, z)$ be the attenuation coefficient distribution of the administered contrast agent. The subtraction operation involves first taking the logs of the measured intensities to derive the desired line integrals and then subtracting. The measured intensities, assuming monoenergetic parallel x-rays are

$$I_1 = I_0 \exp\left[-\int \mu_t(x, y, z)dz \right] \tag{11.9}$$

$$I_2 = I_0 \exp\left[-\int [\mu_t(x, y, z) + \mu_c(x, y, z)]dz \right]. \tag{11.10}$$

Subtracting the logs, we have

$$\ln\left(\frac{I_0}{I_2}\right) - \ln\left(\frac{I_0}{I_1}\right) = \int \mu_c(x, y, z)dz \tag{11.11}$$

the desired projection image of the contrast agent alone.

The quality of this image is limited by two factors, motion and noise. Various physiological motions, often involuntary, occur between I_1, the precontrast image, and I_2, the postcontrast image. This provides a motion "noise" component in the output image given by

$$n_m = \int \mu_t(x, y, z, t_1)dx - \int \mu_t(x, y, z, t_2)dz \tag{11.12}$$

where t_1 and t_2 are the acquisition times of the images. This effect can be minimized by storing a number of precontrast and/or postcontrast images and finding a pair that provides acceptable tissue subtraction performance with a small n_m.

The basic noise sources have been considered in Chapter 6, including Poisson counting noise, additive noise, and scatter. The deterministic scatter component should be essentially identical both before and after the administration of contrast material since the scattering volume is unchanged. Therefore, this scatter component should cancel out, leaving only the statistical noise.

The total Poisson counting noise will therefore be represented by the transmitted and scattered photons of each measurement. Similarly, the additive electrical noise, from devices such as television cameras, will be comparable on each measurement. Since each noise component is independent, the total variance, from equations (6.35) and (6.26) is given by

$$\sigma^2 = \frac{2}{\eta N_t} + \frac{2}{\eta N_s} + 2\sigma_a^2 \qquad (11.13)$$

where N_t is the number of transmitted photons per element, N_s the number of scattered photons per element, and σ_a^2 represents the effective standard deviation of the additive noise. Each component is doubled since the independent variance of the two measurements are added. The resultant SNR, neglecting motion, is given by

$$\text{SNR} = \frac{\sqrt{\eta/2} \int \mu_c dz}{\sqrt{1/N_t + 1/N_s + \eta\sigma_a^2}}. \qquad (11.14)$$

In regions where the anatomy is relatively static, such as the carotid arteries, excellent vessel images have been produced. A representative example is given in Fig. 11.4. The subtraction operation is made highly stable using digital

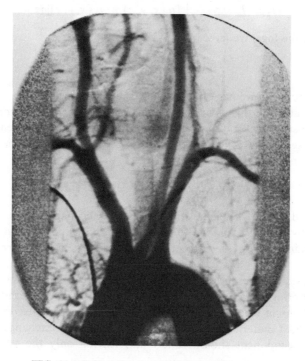

FIG. 11.4 Subtraction image of the carotid artery.

fluoroscopy. The output of the television camera, on each frame, is digitized and stored in a digital memory. The outputs of the two memories are appropriately processed and subtracted to provide the image of Fig. 11.4.

ENERGY-SELECTIVE IMAGING

Since different materials have different energy-dependent attenuation coefficients $\mu(\mathcal{E})$, measurements made at different energies can aid in the identification of specific materials in the body. This process is analogous to the use of color in the visible spectrum, where measurements are made in three spectral regions to enable the identification of the reflectivity of objects by their color.

As indicated in equation (3.13), the attenuation coefficient of materials in the body can be decomposed into a sum of the Rayleigh, photoelectric, and Compton scattering coefficients. As shown in equation (3.14), each individual component can be represented by a constant, weighting a particular function of energy. The energy function is the same for each material, so that each material is completely defined by the weighting constants. If we neglect the relatively small Rayleigh component, the attenuation coefficient of any material can be characterized as [Alvarez and Macovski, 1976]

$$\mu(\mathcal{E}) = \mu_c(\mathcal{E}) + \mu_p(\mathcal{E}) = a_c f_c(\mathcal{E}) + a_p f_p(\mathcal{E}) \tag{11.15}$$

where $f_c(\mathcal{E})$ and $f_p(\mathcal{E})$ are universal functions and a_c and a_p are a pair of constants representing the material. As indicated in Chapter 3, $f_c(\mathcal{E})$ is a rather complex function given in equation (3.16), and $f_p(\mathcal{E})$ is approximately given by $\mathcal{E}^{-3}$.

Equation (11.15) indicates the desirability of a two-dimensional decomposition of x-ray attenuation coefficients. For example, in a computerized tomography system, rather than only making a cross-sectional image of $\mu(x, y)$, additional cross-sectional images can be made of $a_p(x, y)$ and $a_c(x, y)$ which indicate specific material properties. As indicated in equation (3.14), a_c is dependent on the electron density, while a_p is additionally strongly dependent on the atomic number. Thus the pair of processed images delineate the important parameters of average ρ and Z at each pixel.

To create the cross-sectional images a_c and a_p, we must first find their projections or line integrals which we will call A_c and A_p, respectively, where

$$A_p = \int a_p(x, y)dl$$

and $\tag{11.16}$

$$A_c = \int a_c(x, y)dl.$$

These can be calculated by first making measurements at two indpendent energy spectra $S_1(\mathcal{E})$ and $S_2(\mathcal{E})$ as given by

$$I_1 = \int S_1(\mathcal{E}) \exp\left[-\int \mu(\mathcal{E}, x, y)dl\right]d\mathcal{E}$$

and   (11.17)

$$I_2 = \int S_2(\mathcal{E}) \exp\left[-\int \mu(\mathcal{E}, x, y)dl\right]d\mathcal{E}.$$

The two spectra can be obtained by using different anode voltages on the x-ray tubes and/or different x-ray filter material in the beam. Alternatively, energy-selective detectors can be used.

A data-processing system is then required to derive A_p and A_c from I_1 and I_2. Here we make use of the fact that the line integral of the attenuation coefficient is the sum of the photoelectric and Compton line integrals as given by

$$\int \mu(\mathcal{E}, x, y)dl = A_p f_p(\mathcal{E}) + A_c f_c(\mathcal{E}).$$ (11.18)

Substituting (11.18) in the (11.17) measurement equations, we have two equations in two unknowns, A_p and A_c.

If $S_1(\mathcal{E})$ and $S_2(\mathcal{E})$ represented narrow monoenergetic sources, we remove the integration over energy. In that case, by merely taking logs, we have a simple algebraic solution for A_p and A_c. In the general case, however, with broad-spectral sources, we are faced with the solution of nonlinear integral equations. The equations can be solved numerically using power series solutions of the form

$$\ln I_1 = b_0 + b_1 A_p + b_2 A_c + b_3 A_p^2 + b_4 A_c^2 + b_5 A_p A_c + b_6 A_p^3 + b_7 A_c^3$$
$$\ln I_2 = c_0 + c_1 A_p + c_2 A_c + c_3 A_p^2 + c_4 A_c^2 + c_5 A_p A_c + c_6 A_p^3 + c_7 A_c^3.$$ (11.19)

The sets of constants b_i and c_i can be evaluated analytically or preferably with measurements on known materials. In the latter method it becomes preferable to develop values related to the line integral of specific materials themselves, rather than the photoelectric and Compton components. For example, using materials such as aluminum and water, the system can be calibrated for line integrals of aluminum and water A_a and A_w. These represent the actual lengths of the test materials used, resulting in a high degree of accuracy. This decomposition is equivalent to a photoelectric-Compton decomposition and can be easily transformed into the latter with a simple linear transformation [Lehmann et al., 1981].

Once A_p and A_c are determined, or the equivalent based on two actual materials, they can be used to reconstruct cross-sectional images as in conventional computerized tomography (CT) [Alvarez and Macovski, 1976]. For example, using the convolution-back projection algorithm of Chapter 7, images of $a_p(x, y)$ and $a_c(x, y)$ can be reconstructed giving the desired material properties. It is important to note that these images will be free of the nonlinear artifacts discussed in connection with equation (7.69). Since the processed line integrals are energy independent, they are totally free of the nonlinear artifact. Therefore, dual-energy CT systems provide both material information and artifact-free reconstructions.

In addition to computerized tomography, dual-energy reconstructions are very significant in projection radiography. Here dual-energy measurements are acquired as two-dimensional projections of the entire volume. These projections are processed, exactly as previously described to provide two-dimensional image data $A_p(x, y)$ and $A_c(x, y)$. These can be individually displayed to provide material-dependent projection information.

A more exciting presentation is the display of a weighted sum of A_p and A_c [Lehmann et al., 1981], providing an image of the type

$$A(x, y) = A_p(x, y) + rA_c(x, y) \tag{11.20}$$

where r is the ratio of the combined components. This ratio r can be chosen to provide a wide variety of useful clinical images. For example, to eliminate a particular material, r is given by

$$r = -\frac{a_{pi}}{a_{ci}} \tag{11.21}$$

where a_{pi} and a_{ci} are the photoelectric and Compton constants of the material to be removed. This can, for example, be used to eliminate soft tissue or water structures to either display bones, calcifications, or administered iodine-contrast agents.

One excellent example is the intraveneous pyelogram (IVP), where an iodinated contrast agent collects in the kidney. As shown in Fig. 11.5, the conventional IVP normally has intervening bowel gas dispersed among the soft-tissue structures which seriously degrades the visualization of the kidney. Using dual-energy acquisition and processing, a weighted sum image is taken to cancel water or soft tissue. As shown, this eliminates bowel gas interference since it represents variations in soft tissue. The resultant visualization is remarkably improved and has led to the identification of disease which was otherwise missed.

Another important application is tissue look-alike, where any given material can be made to appear like any other. Here the ratio r is chosen as given by

$$r = \frac{a_{p2} - a_{p1}}{a_{c1} - a_{c2}} \tag{11.22}$$

where a_{p1}, a_{c1} and a_{p2}, a_{c2} are the photoelectric and Compton coefficients of the two materials being matched. With this value of r, all equal lengths of the two materials will provide equal values in the resultant image.

An excellent example of this procedure is chest imaging, where important lesions are often obscured by bone. By setting bone to mimic soft tissue, using equation (11.22), the bones will disappear providing an image of the soft tissue only. For example, ribs immersed in soft-tissue structures will have no apparent contrast since they produce the same output as soft tissue. A chest image where the bone mimics soft tissue is shown in Fig. 11.6 together with a conventional chest image. For completeness, Fig. 11.7 shows the same image using the method of equation (11.21) to remove the soft tissue and display the bone.

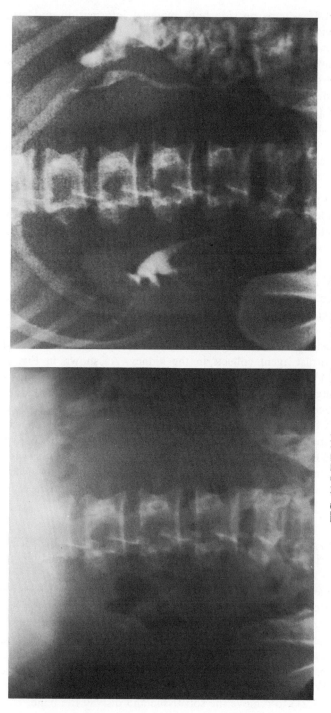

FIG. 11.5 IVP kidney study before and after soft tissue subtraction.

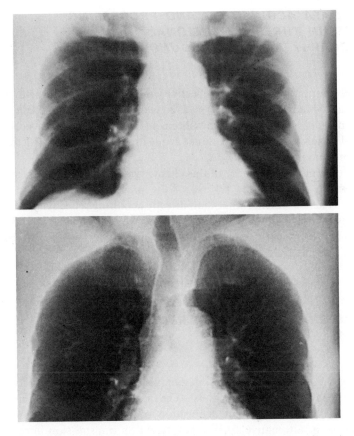

FIG. 11.6 A conventional chest image and a processed image where the bone mimics soft tissue.

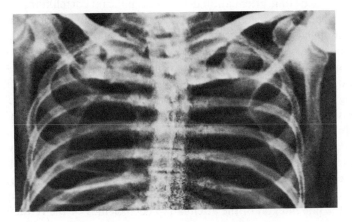

FIG. 11.7 Chest image with the soft tissue eliminated to display the bone.

NOVEL DATA ACQUISITION AND PROCESSING SYSTEMS IN THREE-DIMENSIONAL RADIOGRAPHY AND ULTRASOUND

In general information about the reflectivity of each voxel in a three-dimensional volume can be acquired directly in a conventional ultrasound echo imaging system. In x-ray, however, the measurement of the attenuation coefficient of individual voxels is not done by direct acquisition but by reconstruction of line integral measurements.

Throughout the remainder of this chapter we will briefly discuss speculative systems which represent the reverse of the present approach. These include ultrasonic systems, which, rather than directly acquiring reflectivity information, use reconstruction from line integrals. Also, we will show two radiographic systems where x-ray parameters are acquired directly rather than by reconstruction from line integrals.

LINE-INTEGRAL ULTRASONIC RECONSTRUCTION SYSTEMS

In Chapters 9 and 10 ultrasonic imaging was studied in the reflection mode, the only one in widespread clinical use. The immense popularity of the reflection approach is due to the direct acquisition of three-dimensional information. The impetus for alternative approaches, based on transmission information, is the desire to measure ultrasonic parameters other than reflectivity.

Early results indicate that the localized sonic velocity or refractive index, and the localized sonic attenuation may have significant correlations with disease processes. These parameters are not available in a conventional reflectivity image, although they can be grossly inferred. For example, if the region behind a lesion appears to be echo-free, it may be implied that the lesion exhibits unusually high attenuation, thus reducing echoes behind it.

Reconstruction of velocity and attenuation requires measurement of their line integrals and then a mathematical inversion as described in Chapter 7. The line integrals can be measured by the system shown in Fig. 11.8, which is identical in concept with the x-ray system of Fig. 7.4.

Using a scanned transmitting transducer and a synchronously scanned receiver transducer, a complete set of line integrals are measured at all positions and angles [Greenleaf and Bahn, 1981]. Two measurements are made: the time of flight to measure the line integral of the refractive index and the amplitude to measure the line integral of the attenuation coefficient. For the refractive index,

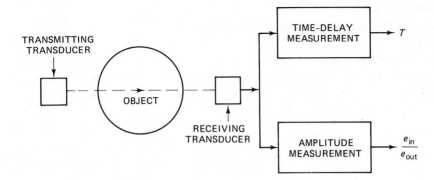

FIG. 11.8 Ultrasonic line-integral data acquisition systems.

this is given by

$$T = \frac{1}{c_0} \int n(x, y)dl \qquad (11.23)$$

where T is the time of flight, c_0 the standard velocity in water, and n the relative refractive index c_0/c. For the attenuation, the line integral is defined as

$$\ln\left(\frac{e_{in}}{e_{out}}\right) = \int \alpha(x, y)dl \qquad (11.24)$$

similar to the x-ray case in Chapter 7. Thus the line integrals of n and α are derived from the measurements.

These measurements are used to reconstruct cross-sectional images of $n(x, y)$ and $\alpha(x, y)$. These have been shown to be diagnostically significant [Greenleaf and Bahn, 1981] in the diagnosis of diseases of the breast. The breast is essentially the only organ where this technique can be applied since it allows for transmission measurements at all angles and positions without intervening air and bone.

Another line-integral approach to ultrasonic imaging has been proposed as an alternative approach to measuring reflectivity [Norton and Linzer, 1979]. If a small transducer is excited with a pulse, it will produce an isotropic pattern represented by expanding circles in a plane. In the receiving mode the output from the transducer, at any given time, represents the line integral of the reflectivity function along a particular circle. The circle is centered at the transducer and has a radius $cT/2$, where T is the round-trip time interval. If a ring of transducers is placed around the object being studied, such as a section of the breast, each transducer will produce an array of measurements of the line integrals of concentric circles. The total array of measurements can be inverted [Norton and Linzer, 1979] to provide a two-dimensional reflectivity function. In this way the reflectivity is measured without attempting to create focused beam patterns aimed at specific regions.

DIRECT ACQUISITION
OF RADIOGRAPHIC PARAMETERS

Classic radiographic transmission measurements represent the line integral of the attenuation coefficient. Systems do exist, however, for directly measuring radiographic parameters. One example is Compton scatter imaging, as illustrated in Fig. 11.9 [Farmer and Collins, 1971].

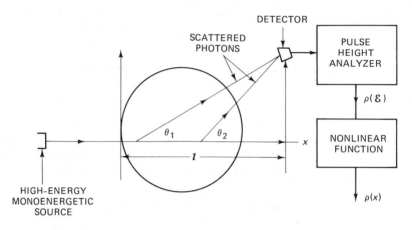

FIG. 11.9 Compton scatter imaging.

A high-energy monoenergetic source is used in the vicinity of 0.5 to 1.0 Mev. This can be derived from radioactive isotopes such as cobalt. The use of relatively high energies ensures a significant energy change during a Compton scattering event. The beam is collimated into a narrow pencil beam and projected through the body. Various Compton scattering events occurring along the path are received by the detector as illustrated in Fig. 11.9.

For a scattering event to reach the detector it must scatter at a unique angle at each x position. As indicated in Chapter 3, the scatter angle is directly related to the energy of the scattered photon. Thus $\theta = f(\mathcal{E})$, as given in equation (3.11). The output of the detector is subject to pulse-height analysis, providing an output of intensity versus energy. This can be converted to intensity as a function of angle using equation (3.11). However, the x position is directly determined by the angle as given by

$$\tan \theta_i = \frac{h}{l - x_i}. \tag{11.25}$$

Therefore, a nonlinear circuit can be used to convert the energy spectrum directly into scattering density as a function of x. The scattering density at each point is directly related to the electron density, which is comparable to the physical density ρ.

Therefore, along the line of the beam, we directly acquire density as a function of position. The beam can be scanned through a plane of interest to provide $p(x, y)$. The system has a few inherent assumptions, such as neglecting multiple scattering events and the attenuation of the beam. Clinical images have been produced.

STIMULATED POSITRON EMISSION

Another recent approach to the direct acquisition of radiographic data is stimulated *positron emission* [Benjamin and Macovski, 1979]. This method is comparable to the positron emission system described in Chapter 8, except that radioactive isotopes are not involved. Instead, a high-energy monoenergetic sheet beam is used to stimulate the emission of positrons with the subsequent annihilation and emission of a pair of 510-kev photons traveling in opposite directions.

Consider the system illustrated in Fig. 8.12. Assume that a sheet beam parallel to the detector arrays and having an energy greater than 1.02 Mev is projected through the object. At this energy [Ter Pogossian, 1967; Johns and Cunningham, 1974] the high-energy photons can give up their energy to form an electron–positron pair. The positron is almost immediately annihilated to produce equal and opposite 0.510-Mev photons, as indicated in Fig. 8.12.

The position of the annihilation region is calculated using the position of the coincident events at the detector as given in equation (8.50). In the case of stimulated positron emission, however, the z position of the planar beam is known, so that the reconstruction is exact. This is in sharp distinction to the isotopic emission case, where the z position of the emitter is unknown and line-integral reconstructions are required.

The resultant planar images represent the attenuation coefficient due to pair production. Since this is proportional to Z^2 [Ter Pogossian, 1967; Johns and Cunningham, 1974], the images emphasize higher-atomic-number materials. They thus can prove useful to image contrast agents or evaluate bone mineralization [Benjamin and Macovski, 1980]. The distinct advantage is the direct acquisition of three-dimensional data, as distinct from computerized tomography. For example, a small region of interest can be studied without making an array of measurements at all angles and positions.

The system has a number of practical difficulties such as requiring energy-selective detectors working at relatively high photon rates. The energy selection is required to distinguish the abundant undesired Compton scattering photons from the desired 0.510-Mev coincident pairs.

References

ALVAREZ, R. E., and MACOVSKI, A. (1976), "Energy-Selective Reconstructions in X-Ray Computerized Tomography," *Phys. Med. Biol.*, Vol. 21, pp. 733–744.

ANDREW, E. R. (1980), "Nuclear Magnetic Resonance Imaging, the Multiple Sensitive Point Method," *IEEE Trans. Nucl. Sci.*, Vol. NS-27, pp. 1232–1238.

ANGER, H. O. (1958), "Scintillation Camera," *Rev. Sci. Instrum.*, Vol. 29, p. 27.

ANGER, H. O. (1964), "Scintillation Camera with Multichannel Collimators," *J. Nucl. Med.*, Vol. 5, p. 515.

BARRETT, H. H. (1972), "Fresnel Zone Plate Imaging in Nuclear Medicine," *J. Nucl. Med.*, Vol. 13, pp. 382–385.

BARRETT, H. H., GAREWAL, G., and WILSON, D. T. (1972), "A Spatially Coded X-Ray Source," *Radiology*, Vol. 104, pp. 429–430.

BATES, C. W., and MORWOOD, R. (1973), "Information Transfer Using Thin Transparent Membranes," *App. Opt.*, Vol. 12, Sept. 1973, pp. 2185–2187.

BENDER, M. A., and BLAU, M. (1963), "The Autofluoroscope," *Nucleonics*, Vol. 21, p. 52.

BENJAMIN, M., and MACOVSKI, A. (1979), "Stimulated Positron Emission (SPE): A New Method of 3-D Tomographic Imaging," *Dig. XII Int. Conf. Med. Biol. Eng.*

BENJAMIN, M., and MACOVSKI, A. (1980), "On the Potential Use of Stimulated Positron Emission (SPE) in the Detection and Monitoring of Some Bone Diseases," *Med. Phys.*, Vol. 7, pp. 112–119.

BLAHD, W. H. (1965), *Nuclear Medicine*, McGraw-Hill Book Company, New York.

BLOCH, F. (1946), "Nuclear Induction," *Phys. Rev.*, Vol. 70, p. 460.

BRACEWELL, R. N. (1965), *The Fourier Transform and Its Application*, McGraw-Hill Book Company, New York.

BROOKS, R. A., and DICHIRO, G. (1976a), "Principles of Computer Assisted Tomography (CAT) in Radiographic and Radioisotopic Imaging," *Phys. Med. Biol.*, Vol. 21, pp. 689–732.

BROOKS, R. A., and DICHIRO, G. (1976b), "Statistical Limitations in X-Ray Reconstructive Tomography," *Med. Phys.*, Vol. 3, pp. 237–240.

BUDINGER, T. F., and GULLBERG, G. T. (1974), "Three-Dimensional Reconstruction in Nuclear Medicine Emission Imaging," *IEEE Trans. Nucl. Sci.*, Vol. NS-21, p. 1.

BURCKHARDT, C. B. (1978), "Speckle in Ultrasound B-Mode Scans," *IEEE Trans. Sonics Ultrasonics*, Vol. SU-25, pp. 1–6.

CHESLER, D. A, RIEDERER, S. J., and PELC, N. J. (1977), "Noise Due to Photon Counting Statistics in Computed X-Ray Tomography," *J. Comput. Assist. Tomogr.*, Vol. 1, pp. 64–74.

CHO, Z. H. (1974), "Special Issue on Physical and Computational Aspects of 3-Dimensional Image Reconstruction," *IEEE Trans. Nucl. Sci.*, Vol. NS-21.

CHRISTENSEN, E. C., CURRY, T. S., and DOWDEY, J. E. (1978), *An Introduction to the Physics of Dianostic Radiology*, 2nd ed., & Febiger, Philadelphia.

CROOKS, L. E. (1980), "Selective Irradiation Line Scan Techniques for NMR Imaging," *IEEE Trans. Nucl. Sci.*, Vol. NS-27, pp. 1239–1245.

DENTON, R. B., FRIEDLANDER, B., and ROCKMORE, A. J. (1979), "Direct Three-Dimensional Image Reconstruction from Divergent Rays," *IEEE Trans. Nucl. Sci.*, Vol. NS-26, pp. 4695–4701.

FARMER, F. T., and COLLINS, M. P. (1971), "A New Approach to the Determination of Anatomical Cross-Sections of the Body by Compton Scattering of Gamma-Rays," *Phys. Med. Biol.*, Vol. 16, pp. 577–586.

FELLER, W. (1957), *An Introduction to Probability Theory and Its Applications*, John Wiley & Sons, New York.

GOODMAN, J. W. (1968), *Introduction to Fourier Optics*, McGraw-Hill Book Company, New York.

GORDON, R. (1975), "Image Processing for 2-D and 3-D Reconstruction from Projections: Theory and Practice in Medicine and the Physical Sciences," *Conf. Proc.*, Optical Society of America, Washington, DC.

GRANT, D. G. (1972), "Tomosynthesis: A Three-Dimensional Radiographic Imaging Technique," *IEEE Trans. Biomed. Eng.*, Vol. BME-19, pp. 20–28.

GRAY, R. M., and MACOVSKI, A. (1976), "Maximum A-Posteriori Estimation of Position in Scintillation Cameras," *IEEE Trans. Nucl. Sci.*, Vol. NS-23, pp. 849–852.

GREENLEAF, J. L., and BAHN, R. C. (1981), "Clinical Imaging with Transmissive Ultrasonic Computerized Tomography," *IEEE Trans. Biomed. Eng.*, Vol. BME-28, pp. 177–185.

GULLBERG, G. T. (1979), "The Reconstruction of Fan-Beam Data by Filtering the Back-Projection," *Comput. Graphics Image Process.*, Vol. 10, pp. 30–47.

HERMAN, G. T. (1980), *Image Reconstruction from Projections*, Academic Press, New York.

HORN, B. K. P. (1978), "Density Reconstruction Using Arbitrary Ray-Sampling Schemes," *Proc. IEEE*, Vol. 66, pp. 551–562.

JOHNS, E. J., and CUNNINGHAM, J. R. (1974), *The Physics of Radiology*, Charles C Thomas, Springfield, IL.

KLEIN, O., and NISHINA, Y. (1929), "Über die Streung von Quantendynamik von Dirac," *Z. Phys.*, No. 52, p. 853.

LARSEN, L. E., and JACOBI, H. (1978), "Microwave Interrogation of Detective Targets, Part I: By Scattering Parameters," *Med. Phys.*, Vol. 5, No. 6, pp. 500–508.

LAUGHLIN, J. S., BEATTIE, J. W., COREY, K. R., ISAACSON, A., and KENNY, P. J. (1960), "A Total Body Scanner for High Energy Gamma Rays," *Radiology*, Vol. 74, p. 108.

LAUTERBUR, P. C., and LAI, C. M. (1980), "Zeugmatography by Reconstruction from Projections," *IEEE Trans. Nucl. Sci.*, Vol. NS-2, pp. 1227–1231.

LEDLEY, R. S. (1976), "Special Issue: Advances in Picture Reconstruction Theory and Applications," *Comput. Biol. Med.*, Vol. 6.

LEHMANN, L. A., ALVAREZ, R. E., MACOVSKI, A., PELC, N. J., RIEDERER, S. J., HALL, A. J., and BRODY, W. R. (1981), "Generalized Image Combinations in Dual KVP Digital Radiography," *Med. Phys.*, Vol. 8, No. 5.

MACOVSKI, A. (1979), "Ultrasonic Imaging Using Arrays," *Proc. IEEE*, Vol. 67, pp. 484–495.

MACOVSKI, A., and NORTON, S. J. (1975), "High Resolution B Scans Using a Circular Array," *Acoustical Holography*, Vol. 6 (N. Booth, ed.), Plenum Press, New York, pp. 121–143.

MCLEAN, T. P., and SCHAGEN, P. (1979), *Electronic Imaging*, Academic Press, London.

MEREDITH, W. J., and MASSEY, J. B. (1977), *Fundamental Physics of Radiology*, John Wright & Sons, Bristol, England.

METZ, C. E., ATKINS, F. B., and BECK, R. N. (1980), "The Geometric Transfer Function Component for Scintillation Camera Collimators with Straight Parallel Holes," *Phys. Med. Biol.*, Vol. 25, pp. 1059–1070.

NICHOLAS, D. (1977), "An Introduction to the Theory of Acoustic Scattering by Biological Tissue," in *Recent Advances in Ultrasound in Biomedicine* (D. N. White, ed.), Research Studies Press, Forest Grove, OR, Chap. 1.

NORTON, S. J. (1976), "Theory of Acoustic Imaging," Ph.D. dissertation, Dept. of Electrical Engineering, SEL Tech. Rep. No. 4956-2, Stanford University, Stanford, CA.

NORTON, S. J., and LINZER, M. (1979), "Ultrasonic Reflectivity Tomography: Reconstruction with Circular Transducer Arrays," *Ultrason. Imaging*, Vol. 1, pp. 154–184.

OVITT, T. W., CAPP, P. M., FISHER, D. H., FROST, M. M., LEBEL, J. L., NUDLEMAN, S., and ROEHRIG, K. (1978), "The Development of a Digital Video Subtraction System for Intravenous Angiography," *Proc. SPIE*, Vol. 167, pp. 61–66.

PAPOULIS, A. (1965), *Random Variables and Stochastic Processes*, McGraw-Hill Book Company, New York.

PARZEN, E. (1960), *Modern Probability Theory and Its Applications*, John Wiley & Sons, New York.

SCUDDER, H. J. (1978), "Introduction of Computer Aided Tomography," *Proc. IEEE*, Vol. 66, pp. 628–637.

SOMER, J. C. (1968), "Electronic Sector Scanning for Ultrasonic Diagnosis," *Ultrasonics*, Vol. 6, pp. 153–159.

SORENSON, J. A., and NELSON, J. A. (1976), "Investigations of Moving Slit Radiography," *Radiology*, Vol. 120, pp. 705–711.

SPRAWLS, P. S., JR. (1977), *The Physical Principles of Diagnostic Radiology*, University Park Press, Baltimore, MD.

STONESTROM, J. P., ALVAREZ, R. E., and MACOVSKI, A. (1981), "A Framework for Spectral Artifact Corrections in X-ray CT," *IEEE Trans. Biomed. Eng.*, Vol. BME-28, pp. 128–141.

TANAKA, E. (1979), "Generalized Correction Functions for Convolutional Techniques in Three-Dimensional Image Reconstruction," *Phys. Med. Biol.*, Vol. 24, pp. 157–161.

TER-POGOSSIAN, M. M. (1967), *The Physical Aspects of Diagnostic Radiology*, Harper & Row, Inc., Hagerstown, MD.

TER-POGOSSIAN, M. M., PHELPS, M. E., HOFFMAN, M. E., and MULLANI, N. A. (1975), "A Positron-Emission Transaxial Tomograph for Nuclear Imaging (PETT)," *Radiology*, Vol. 114, pp. 89–98.

WEIDNER, R. T., and SELLS, R. L. (1960), *Elementary Modern Physics*, Allyn and Bacon, Inc., Boston.

WELLS, P. N. T. (1969), *Physical Principles of Ultrasonic Diagnosis*, Academic Press, London.

WOODCOCK, J. P. (1979), *Ultransonics (Medical Physics Handbooks: 1)*, Adam Hilger Ltd., Bristol, England.

Index

noise

Poisson
statistics

$$\frac{\sqrt{Bgd. * \frac{photons}{pixel}}}{Signal}$$